This Sacred Earth

Scientific and Religious Perspectives on Nature and Humanity's Place Within It

Edited by Paul J. Kirbas

This Sacred Earth

Copyright ©2011

ISBN: 978-1-55605-439-6

Ebook: 978-1-55605-440-2

Library of Congress Control Number: 2011939340

WYNDHAM HALL PRESS
5050 Kerr Rd
Lima, Ohio 45806
www.wyndhamhallpress.com

Printed in The United States of America

TABLE OF CONTENTS

Introduction

From the Editor...

A quick survey of new book titles, periodical covers, and websites reveals that there is a large and growing attention to the issues of creation care, environmental ethics, and global warming. Indeed, it is becoming very fashionable for all sectors of society to address the environment, and to promote the "greening" of their businesses. Although this growing consensus of concern for the environment is positive and encouraging, it nevertheless seems to be a bit, shall we say, shallow and ineffective. I say "shallow", because as I have surveyed a sampling of these expressed environmental concerns, the emphasis seems to be largely on preserving nature as a resource for humans. We are making sure that we have all the comforts that we currently enjoy, and making sure that we pass on a usable earth to our children and grandchildren. This seems to be the motivating factor for our environmental concerns. In this approach, nature takes on a utilitarian function. What is important to us is not necessarily nature. What is important to us is us.

A self-centered concern for nature is a concern, no doubt, but not one with depth; not one rooted in the value of nature itself. For that reason, the concern is largely ineffective. To be certain, shining examples of caring for nature do exist. For the most part, however, there is a disconnection between what people believe about the environment and what they do in response to that belief.

Perhaps there is another approach that should be taken. Rather than seeing nature as important due to its resourcefulness for human activity, is it possible to envision nature as important and valuable in and of itself. Can we claim nature as sacred? And if so, can we envision a proper place for humanity within that sacred nature?

The exploration of the sanctity of nature is not the purview of any one discipline. Certainly theology would want to speak to this, but science should be in the conversation as well. Science and religion are often portrayed as rivals. Yet, in exploring the sanctity of nature, science needs religion, and religion needs science. Theology can teach us the meaning of sacredness, and science can lead us to a better understanding of the scope and the origins of nature. As we focus our attention on the human place in nature, other disciplines are also needed. Philosophy, rhetoric, and human psychology can all help inform this important conversation. Bringing leading thinkers from all of these fields together to reflect on the Sanctity of Nature is an important undertaking. That is what we did, and this book offers the first results.

In the fall of 2010, The Kirbas Institute (www.kirbasinstitute .org) and the Fetzer Institute (www.fetzer.org) jointly hosted a 3 day dialogue on Fetzer's campus in Kalamazoo, Michigan. We invited leading thinkers from the fields of theology, the sciences, and liberal arts. Our guests came from leading universities and organizations from around the world. Together, we explored the concept of the sanctity of nature, and of our human place within it. This book offers a collection of the contributions that each participant presented as a part of our dialogue. Between the essays of this book, I have created an editorial bridge that demonstrates the progression of thought that emerged in our dialogue. The Teachers' Edition of this book also contains a DVD featuring video interviews with our participants.

As we begin this journey of exploring the sanctity of nature, perhaps it is most appropriate to step back, disengage heavy intellectual pursuit, and simply observe and contemplate. Before dissecting the sanctity of nature, let's experience it. Let's begin with awe, as we peer into the incredible universe of which we are but a small part. Our tour guide for this first chapter is Astronomer Jennifer Wiseman.

Dr. Jennifer Wiseman has studied star-forming regions of our galaxy using radio, optical, and infrared telescopes, and currently serves as Chief of the Laboratory for Exoplanets and Stellar Astrophysics at NASA Goddard Space Flight Center. She studied physics at MIT for her bachelors degree, discovering comet Wiseman-Skiff in 1987. She then earned a Ph.D. in astronomy at Harvard University in 1995. She accomplished subsequent research as a Jansky Fellow at the National Radio Astronomy Observatory and as a Hubble Fellow at the Johns Hopkins University. In addition to research, Dr. Wiseman is also interested in public science policy. She was selected as the 2001-2002 Congressional Science Fellow of the American Physical Society, and served on the staff of the Science Committee of the U.S. House of Representatives. She then served from 2003-2006 as the Program Scientist for the Hubble Space Telescope at NASA Headquarters in Washington, DC. Dr. Wiseman has also authored several essays addressing the relationship of astronomy and Christian faith.

The following chapter is a summary of the verbal presentation that Dr. Wiseman offered as the initial presentation of our gathering at the Fetzer Institute.

Chapter 1
Beginning With Awe
Peering into the Universe

A Summary of a Presentation
Offered by Jennifer Wiseman

Prepared by Paul J. Kirbas

On a cool Friday evening during the fall season, scientists, theologians, philosophers, and others assembled from various parts of the world and gathered on the campus of the Fetzer Institute in Kalamazoo, Michigan. We were beginning a weekend of dialogue on the important theme of the sanctity of nature, the fruits of which comprise this book. Each participant was prepared with his or her own writings, and we had many deep and detailed ideas to work through. Given the fact that many of our participants had just arrived from long and arduous travels from international locales, there was a prevailing sense of tiredness among us all. In our initial planning, we had decided that for that first night, we would simply provide a welcome dinner, and send our participants off for a good night's rest. We later decided that the first evening should have one presentation included, yet one that would offer an inspirational focus; One that would energize our participants, and help them find a sense of grandeur and significance in our collective enterprise. We wanted to begin *with awe*. We realized that we had just the right person to lead us: Astronomer Jennifer Wiseman. On that first evening of our gathering, following a welcome dinner, Dr. Wiseman took the podium, and gave us a guided tour of this amazing universe in which we live.

Dr. Wiseman began by sharing a bit about her own story, and told us of her earliest experiences of sensing awe for the universe. She grew up on a cattle farm, with a very down to earth family. Yet her down to earth family helped her begin the journey of becoming a star gazer. Her love of space began as her family took walks on the dark farmland at night. She recalls gazing up at the night sky, seeing the brilliant display of stars from one horizon to the other. She was awed and fascinated by this sight. This was also the era of Carl Sagan and his *Cosmos* series. As a young person, she was filled with excitement about pictures that came back from the early probes that were sent into space. She recalls thinking that this was the most amazing thing that humans could ever do. To this day, she still feels this way.

With images on the screen and a wireless mouse in her hand, Dr. Wiseman offered us a tour of our universe. She began by showing us a beautiful picture of a spiral galaxy. She explained that galaxies such as this are full of gas, dust, and stars that could number in the billions. Our planet exists within one of these spiral galaxies, but we cannot get outside of our own to see it. We are able to get a sense of it from our internal vantage point, as well as from our ability to see comparable galaxies beyond our own. Interestingly, it was only determined in recent decades that there were galaxies outside our own.

As she led us on this tour of the universe, Dr. Wiseman introduced us to the Hubble telescope, one of the main tools of our exploration. The Hubble telescope is in orbit around the earth, making a complete circle around our planet every ninety minutes. Dr. Wiseman's current role at NASA is as a Senior Project Scientist for the Hubble Telescope. Hubble is designed to accommodate Space Shuttle dockings so that astronauts can periodically perform service missions. These missions have greatly enhanced Hubble's power to look deep into our universe. In order to gain an appreciation for how much technology has enhanced our ability to look into the universe, Dr. Wiseman presented a chart that shows the increase in increments of the sensitivity of what we are able to see, starting from the astronomical observations of Galileo some 400 years ago to Hubble's ability today. The most recent servicing mission has pushed the sensitivity abilities off the chart. In Hubble today, we have a truly state-of-the-art instrument, more powerful than ever before.

Our observations of the universe, Dr. Wiseman suggested, lead us to some clear perspectives that we should take in and reflect upon. These perspectives are that of magnitude, beauty, activity, and faith.

Perspective of Magnitude

One thing that becomes obvious, albeit incomprehensible, is the issue of the magnitude of the universe-both in terms of the size of space and the numbers of objects that inhabit it. This was illustrated by a picture of the Sagittarius Star Cloud, which is in our own galaxy. Upon seeing it, most of us were immediately struck by three things: the diverse colors of the stars; the different degrees of size and brightness of the stars; and the sheer number of stars. And to think, this is in one particular section of our particular galaxy. Today, we have some sense of how huge galaxies such as ours can be. Our galaxy is 150,000 light years across. One light year is 10 million million kilometers. We believe that our sun is located about two thirds of the way out on one of the spiral arms. The magnitude of our own galaxy is mind boggling, yet the distances between galaxies are even more so. Our closest "neighboring" galaxy is the Andromeda Galaxy, which is 2 million light years away. This means that as Hubble sees the light of Andromeda, it is really seeing it as it was 2 million years ago. Through Hubble's amazing sensitivity, we are now seeing the light from galaxies that are several billion light years away, meaning that we are seeing them as they were several billion years ago. We believe, according to Dr. Wiseman, that we are currently seeing things that are nearly as old as the universe itself. With this expansive view, it is now believed that our universe contains 100 billion galaxies, with 100 billion stars in each one. Earth is one planet orbiting one of those stars in one of those galaxies. Indeed, peering into the universe gives us a perspective of magnitude that leaves us in true awe, and keeps us very humble when we consider our place within it.

Perspective of Beauty

As we viewed the image of the Sagittarius Star Cloud, we noted another observation: the beauty of the colors. Dr. Wiseman told of a student who picked a random galaxy, NGC1309, and was truly inspired by its beautiful symmetrical shape. The student created an art piece of this image, stating that when she looked at the spiral galaxies, they seemed to be dancing. Indeed, there is tremendous beauty in space. To demonstrate this, Dr. Wiseman showed us a picture of Orion. While this scene is beautiful on its own, when seen by Hubble's powerful telescope, one can see beautiful images between the stars, swirls and swirls of ionized gas that is present just after stars are formed, before they begin to come together. This offers a true sense of beauty in the universe, but not beauty that is random. This beauty illustrates a true sense of progression. Consider the following example: In one of Hubble's images, one can see galaxies both nearby and far away, yet at the same time. The galaxies that are the farthest away are moving the fastest due to the expansion of the universe. They appear to have a more red color. These galaxies represent the oldest ones, and as they are compared to the closer, newer galaxies, one can see an amazing trend. In the earlier, older galaxies, the colors indicate that what we see is mostly hydrogen and helium. As we come closer to our epic in time, we see a much richer combination of these elements, plus others such as carbon and iron oxygen. We can use astronomy as a kind of time machine to compare earlier galaxies in the universe with later ones, to see what has been happening over time. In this way, we see true beauty in the diversity of colors, shapes, and sizes. This beauty is striking for its own sake, but also gives us important insights into the secrets of our universe.

Perspective of Activity

In this tour of the universe, one thing that has become apparent is that the universe is not stagnant. Many people have the false sense that it is. We stare into the night sky, thinking that everything we see is frozen in space and time. Actually, there is quite a lot going on. Galaxies are on the move, and if they are close enough to each other, will eventually merge. Dr. Wiseman showed an image of two galaxies coming together. In many ways, it is a bad scene. There can be tremendous havoc. These colliding galaxies lose their spirals, and the resulting turbulence causes a huge surge of new star formation. Bright regions show where new stars are being formed. The resulting image was striking. The core of one galaxy was stripped of its spiral arms and situated in the middle of another galaxy. As Dr. Wiseman was showing this progression I was recalling that our solar system is on the spiral arm of our galaxy, so if this was our collision, we would not have fared so well!

So far in this tour of the universe, we get the sense that we know a lot about our universe. Indeed, our knowledge has grown tremendously over time. Yet, there is much more mystery than there is knowledge. One thing that has become apparent to us is that the expansion of the universe, which we have long thought would show signs of slowing down, is actually accelerating. Although there must be some force acting upon the universe to cause this acceleration, we do not know what it is. We call it "dark energy". Another mystery of the universe is that within galaxies, there is some kind of matter present that we cannot see, but we know it is there because we can observe its gravitational effect on other objects. We call this "dark matter".

If we think of the universe as a "budget of stuff", then, we can realize that in terms of this budgetary basket, we don't understand much. Dark energy, which we don't understand, represents about 73%; Dark matter, which we don't understand, represents about 22%. What we do understand comprises about 4% of the budget. Obviously, notwithstanding all of our amazing achievements in recent decades, we must maintain a healthy humility.

And in terms of the mystery of the universe, and its activity, one of the biggest questions intriguing us is whether there is another earth-like planet somewhere, and if so, does life exists somewhere else in the universe? Dr. Wiseman informed us that we are making great strides in the study of extra solar planets in the last 18 years. To date, we have discovered over 400 of them. These are planets orbiting other stars. They have all been detected by indirect means, by observing the motion of stars. What astronomers would really like to do is to get a picture of one of these exoplanets, but a planet is far dimmer than the parent star so getting such a picture is very hard to do. Dr. Wiseman is hopeful, and states that we are getting closer.

Indeed, there is much going on in that vast universe that we peer into in a dark night's sky. It may seem stagnant to us, but the universe is teeming with activity, only a small portion of which we actually understand.

Perspective of Faith

Can our observations of the universe tell us anything about the spiritual realm, and about the nature of God? Theologians from centuries ago thought that astronomy might tell us something of theological importance. John Calvin said that "astronomy is not only pleasant; it's also useful to know. It cannot be denied that this art unfolds the wisdom of God" But one does not have to see it that way. There are fine scientists who come to a different conclusion, as illustrated by Steven Weinberg who said "the more the universe seems comprehensible, the more it also seems pointless". There is not a prescribed direction that the study of the universe can take a person in terms of one's personal interpretation, philosophical or religious. One thing we need to remember is that these realms (science and religion) answer different types of questions. Science addresses questions of how, when, and why (in terms of physical cause and effect). Faith questions ask the bigger Why. Is there some bigger purpose? Is there a *Who*, some being behind it?

Jumping from issues that science can address to issues that science cannot address, if one does have faith in a divine creator, Dr. Wiseman believes that we can learn things about the character of the Creator through astronomical observations. The following statements are her personal reflections on what one could infer from the character of the universe:

> *This creator appears to be powerful and creative.*
>
> *This creator loves beauty.* There seems to be a connection between the beauty in creation and the existence of beings in creation (humans) who can recognize that beauty.
>
> *This creator is patient.* For us, 13.7 billion years is a long time. Why not just snap your fingers and have things in final form instead of this continuing evolutionary track?
>
> *This creator is faithful.* This universe observes physical laws. There is true order to it.
>
> *This creator is providential.* We live in a fruitful universe. We don't fully comprehend how fruitful the universe may be, because we don't know how many other planets are out there where inhabitants are having the same discussions.

Dr. Wiseman observed that these inferences about God, gleaned from our observation of the universe, are very similar to the words of Scripture, especially the Psalms. In these passages, the meticulous details of how God did this or that are not discussed, but rather, nature is lifted up in praise, invoking a sense of awe. In her own work, Dr. Wiseman collaborates with scientists all over the world. Although they may not all share the same religious beliefs, they can all look up at the sky and be awestruck by its beauty.

Dr. Wiseman is, herself, a Christian. She believes that within the Christian domain, there is another depth to this discussion. We must bring in the person of Jesus Christ. For many Christians, this relevance is confused. They learn one thing in church (a strong focus on Jesus Christ) and in the world, vague discussions about science and God, but they don't seen how the two relate. In Christian scripture, the relationship between the spiritual and the physical is in the person of Jesus. The Bible teaches us that the universe is the product of this Christos, this living word. In the Gospel of John, there are beautifully poetic passages about the beginning. We are told there that the word became flesh, and that while the universe was made through him, he made his dwelling place among us. This Word was not some afterthought. The New Testament book of Hebrews tells us that it was through this Son that God made the whole universe and is, through him, sustaining all things. Dr. Wiseman believes that this is something that does not get enough attention in our churches; that the center of our Sunday morning worship is also the one who is sustaining the universe.

In terms of application to our ethics as Christians, she stated, "So, how should we respond? We feel wonder and awe; we give praise; and we reflect on our place in the universe. We should tend our planet, and we need to be caring for the earth and all of its inhabitants."

In closing, Dr. Wiseman showed us one more image. It was a picture of earth, taken from the vantage point of Saturn. When the spacecraft orbiting Saturn came around the planet, it happened to be in the right position to see earth, a tiny speck in the far off distance. It is quite astounding to see your own planet from such a vantage point. This small fragile planet is our home.

Dr. Wiseman concluded her remarks by stating her hopes that our dialogue would bring about ways in which our spirituality and values are harmonized with what we learn from science in a way that helps people to work together to make our planet a better place.

From the Editor....

As Jennifer Wiseman so aptly points out, it is very difficult to peer into the universe, of which our planet is such a small part, without feeling some sense of awe deep within our hearts. This sense of awe is felt by people of all walks of life, of all faith traditions, and of all perspectives. The awe invocated by peering into the universe may be a common experience for many, but we tend to name that feeling, and describe it to others, with images and symbols that are particular to our own traditions, mindsets, and world views. For Dr. Wiseman, someone who believes in God, this sense of awe is immediately translated into an act of worship, and into a sense of discovery into the nature of the Creator God. As she solidifies this awe into a theistic expression, she finds commonality with all who believe in a creator God: Jews, Muslims, Christians, and many others. Dr. Wiseman further defines this sense of awe, when peering into the universe, in the particular expression of a Christian. As such, she continues to find commonality with other Christians, but her description of the feeling of awe begins to part ways with the others. Deeper into her own particularity, we find that Dr. Wiseman ascribes to the scientific narrative of how God created this awe-invoking universe. She believes in a universe that is billions of years old, and one that has evolved in the ways that science has expressed. In describing the origin of this universe within an evolutionary theory, her views begin to part ways with other evangelical Christians, who maintain a commitment to the literal reading of the Bible, and the Genesis account of creation.

In the journey of demarcations described above, the central and core experience, that of feeling a sense of awe when peering into the universe, is equally experienced by all; yet when we describe that experience, we find ourselves fracturing into many different camps.

Once we find ourselves in these camps, we lose our ability to find common ground, even on the initial and universal sense of awe with which we all start. This is our current challenge in this book. We are attempting to find some common ground on what it means to see nature as sacred, and what our human place and role is within this sacred nature. In order to develop a roadmap to these goals, we will need to begin to clothe this basic sense of awe in certain particular ideals. We can't discover what our human place is within nature, for instance, without identifying how nature got here, and how we arrived within it. We can't understand our human relationship to nature, without identifying the nature of what it means to be human. It is necessary, then, to adopt a certain narrative; to embrace a certain rhetoric.

Before going forward, then, it is important for us to understand the power of narrative and rhetoric, as well as the challenges that we find within that power. Our next essay, then, is about rhetoric. We have invited Dr. David Thomas to write this essay. Thomas has been a distinguished scholar and professor in the fields of Rhetoric and Debate. He has held teaching positions at Concordia College, Auburn University, the University of Houston, and the University of Richmond. In the following essay, Dr. Thomas begins with a general overview of the field of rhetoric. Following that, he applies the tools of rhetoric study to the passionate debate concerning origin theories, such as literal creation and evolution.

Chapter 2
Talking to Fish about Water:
Rhetorical Perspectives towards
Bridging the Divisions
Between Faith and Science

by David A Thomas

Among some scientists, theologians, and philosophers, the whole idea of rhetoric is still an object of general distrust. "Mere" rhetoric is rejected out of hand as consisting of purple prose, political hot air, or worse, deliberate lies and obfuscations, by anyone who fails to engage their ideas in an acceptable dialectical manner. As a rhetorician myself, I admit that when rhetoric meets those pejorative definitions, it ought to be rejected out of hand. (The same may also be said for "mere" science, theology, and philosophy, whenever they are equally open to charges of superficiality, irrationality, and even unethical abuses. But such violations are much easier to identify without necessarily condemning the whole enterprise of science, theology, or philosophy per se.)

At its best, understanding rhetoric is vital and essential to rational deliberation and decision making. It is too bad that scientists, theologians, and philosophers – the specialists in the fields most concerned with questions and issues surrounding the Sanctity of Nature – lack much in the way of rhetorical education. My essay is aimed at unfolding the unique insights into the subject that rhetoric affords. In order to do that, I perceive that I must first lay a basic groundwork. The first part of my essay deals with basic rhetorical theory, both its Classical roots, and its most exciting, relevant modern advances during the past century. The second part of my essay applies these principles of rhetoric to the scientific and theological subjects of most interest to the other contributors to this volume of essays on the Sanctity of Nature.

How often do you hear, "Skip the rhetoric and get down to the facts," or, "We want action, not empty [or mere] rhetoric." I have grown accustomed to typical statements like this one in Denis Alexander's, *Creation or Evolution,* where he wrote, "But if we investigate beyond the rhetoric, does evolutionary theory really stand the weight of such ideological investment?"[1] On one hand, there's *rhetoric* about evolution (what *others* say)*,* then again on the other hand, there's *what is really true* about evolution (what *we* say). In part, the denigration of rhetoric is simply a matter of common usage. As a rhetorician, I question the legitimacy of such a pejorative anti-rhetorical attitude. If I don't defend rhetoric, who will?

My question is, when people of honesty, competence, and integrity do not agree, how can it help to bypass rhetoric? What is the alternative? To the outside auditor/reader, which advocate is justified in claiming to possess Truth for certain? At times, certainty might be justified; maybe there is an unanswerable case. I suspect that on some of the issues we have before us in this book, folks of good will and good credibility will not all agree with each other. Theologians, philosophers, and scientists alike, may be tempted to express themselves with certitude in the heat of the ongoing, *evolving*, [deliberate choice of word] conflicts between faith and science, in the creation/evolution arena, as an example. It is not always clear who is entitled to the presumption of absolute Truth, and who are just ignorant, unreasonable, or dishonest.

We all use rhetoric. When you use language to discuss evolution, when you use a figure of speech to make something about it clear, and when you show a picture to illustrate it, then you, too, are using rhetoric. If consensus and not coercion is the goal, rhetoric is required. As one of my friends suggested, if you have strong convictions against rhetoric, try to get along without it.

FOUNDATIONS: Selective Overview of CLASSICAL Rhetoric

Rhetoric figures in the legacy of classical literature come into Western education from the ancient Greeks, especially the writings of Plato and Aristotle. Plato, the philosopher, [*philo,* lover + *sophis,* wisdom or learning] searched for the ideal forms of Truth, Beauty, and Justice. He asserted in The Republic that only philosophers should be kings. Plato's star pupil, Aristotle, was more of a democrat. Aristotle wrote *The Art of Rhetoric* [*rhema,* word + *techne,* practical art; "art of the orator"] as a handbook for training citizens to be

19

persuasive in their public advocacy. Hence, Aristotle's *The Art of Rhetoric* is part theory, and part prescriptive rules. Concerning the value of rhetoric, Plato taught that rhetoric stands in the same relation to Justice, as cookery (quackery) does to medicine *(Gorgias).*[2]

Tension between philosophy and rhetoric still endures, perhaps unfortunately, perhaps justifiably, into the present. Some philosophers, like Nancey Murphy, are an exception to the rule. In her book, *Reasoning and Rhetoric in Religion,* she favorably discusses Aristotle's classical principles.[3] I have read some rhetorical theorists who are also philosophers. In fact, a decent journal exists that is entitled *Philosophy and Rhetoric.*

Rhetoric merits an honored position in the history of Western education. Medieval universities, like Oxford and Cambridge, taught the Trivium of rhetoric, logic [or dialectic], and grammar, together with the Quadrivium of arithmetic, geometry, music, and astronomy. Those who completed these seven arts and sciences earned the Master of Arts degree. Thereupon, one was prepared to pursue higher education towards a doctorate in the more esteemed areas of philosophy or theology. Mastery of rhetoric, along with dialectic, math, and science, was prerequisite to becoming a philosopher or theologian.[4] In the academy, if you wish to pursue a classical education, you can still study traditional rhetoric in depth.

PRIMER IN ARISOTELIAN CLASSICAL RHETORIC.

Aristotle's *Art of Rhetoric* defines rhetoric as, "The art of discovering, in the given case, what are the available means of persuasion."[5] Thus, rhetoric is an academic discipline, whose object of study is persuasion. Aristotle never denied the value of philosophy. Case in point: *Nicomachean Ethics.* However, to Aristotle, politics, not philosophy, was the highest goal of education for citizenship. To help the citizenry learn to govern themselves instead of being ruled by self-appointed (dare we say *elitist)* philosopher-kings, Aristotle taught the practical art of using oratory to persuade people in certain types of public situations.

Aristotle applied his rhetoric to three main places where public oratory was relevant in Athens of his day: *forensic* (the courts, where guilt or innocence was adjudicated by juries), *deliberative* (the forum or senate where legislation was argued and voted upon by the assembly), and *epideictic* (ritual oratory on ceremonial occasions, like the speaking events at the Olympic games).

Each of these three ancient oratorical venues had its particular characteristics owing to their distinctive purposes. Forensic oratory dealt with *past fact:* was the contested issue a crime or a tort?; was the accused guilty as charged? Let both sides present their cases, and put it to the jury to decide. Court cases are, at bottom, factual issues to settle.

Deliberative oratory dealt with *future policy*; what should be done about pressing pragmatic issues like war and peace, domestic welfare, and taxation, etc. As was true for forensic cases, facts are the basis for proposed policies, only, as the facts are interpreted in accord with how they are to be *valued* by the will of the people. Policy deliberation is the give and take, airing the pros and cons, engaging in arguments over what facts are, which facts count, and how much. Is a change in policy needed? Would a change be feasible? What are the costs of action vs. inaction? Would a change be beneficial? In view of competing interests, and inherent uncertainties, decisions must usually go forward on the basis of probabilities rather than certainty. Facts alone do not drive public deliberations; facts must be weighed in accord with competing values, as argued in the deliberative process. In addition, in Ancient Greece, as in modern politics, financial and political interests come into play in the forum debates. In the end, deliberative rhetoric deals with matters of expediency, and it always involves risk factors.

The third place for Classical rhetoric, epideictic oratory, calls for speeches of *praise or blame to the gods*, i.e. speeches of inspiration, pride, and national identity. The best familiar analogies for epideictic speaking are patriotic Fourth of July celebrations, eulogies, nominating speeches, dedications, and a lot of Sunday morning sermonizing. Epideictic speaking is character- and value-driven. In today's educational context, training in rhetoric focuses more on forensic and deliberative modes of rhetoric, together with logical standards and tests of evidence. Epideictic rhetoric is for promulgating basic values, and transmitting cultural narratives. I want to mention one more keen Aristotelian insight: when his definition of rhetoric mentioned "the *available means of persuasion*," he had three specific kinds of proofs in mind. These were *ethos, pathos, and logos,* or, ethical proof, emotional proof, and logical proof. In fairness to Aristotle's true brilliance, these three means of persuasion were more broadly construed than our modern, simplified English-translation tags would indicate. *Ethos* denoted the overall credibility of the speaker, based on honesty, competence (knowledge), and his good will towards the audience.

Aristotle regarded *ethos* as the highest form of persuasion. *Pathos*, the root of our English words *empathy and sympathy*, meant more than just cheap manipulation of people's basest emotions; rather, it got at the heart of discerning, and appealing to, people's deepest motivations and drives. Consider *pathos* to be audience psychology – using fear appeals, or more positively, altruism, for example.

Finally, *Logos* does include logical content — reasoning, evidence, and organization. Beyond that, the word *logos* can encompass more than left brained appeals. Bible scholars will recognize that *logos* is the word used in the first chapter of the Gospel of John, "In the beginning was the *logos,* and the *logos* was with God, and the *logos* was God." In classical Greek, the meaning of *logos* also includes *mythos,* storytelling, along with hard logic, a fact that easily gets lost in our contemporary curriculum.

Intellectual Respectability of Classical Rhetoric.

Fast forward to the present, classical rhetoric is still taught in many places as the art of public discourse by a speaker (or writer) for the purpose of persuading audiences. These tenets of Aristotle's theory, if we can call his book of instructions a theory, frequently appear in required courses implicitly even when credit is not exactly attributed to him. Our contemporary use of the principles of *The Art of Rhetoric* begins with public speaking in general. As mentioned, rhetoric may be either oral or written, and English departments also incorporate the principles of rhetoric in writing composition textbooks. Beyond the basic courses, rhetoric may also be listed under other titles like public relations, advertising, or campaigning, jury pleading, and homiletics. It also includes nonverbal and visual rhetoric, as well as verbal rhetoric. For the purpose of expanding the definition and staking out broader disciplinary boundaries, "The given cases" in Aristotle's definition of rhetoric's relevant arenas now incorporate communication studies in all of the different media, including the internet.

PRIMER ON KENNETH BURKE:
Recent Developments in the Twentieth Century.

Today, I identify myself as a narrative rhetorician, instead of the argumentation specialist I started out to be in my youthful debate coaching career. What I am now, I owe to the influence of Kenneth Burke and Walt Fisher.

Kenneth Burke (1897-1993) was a prolific American literary theorist and philosopher. Burke's innovative ideas are considered by many to be the most important advances in rhetoric since the Classical Greek and Latin sources.

1. Burke's Dramatism.

The aim of rhetoric is to induce cooperation through a process of *identification* rather than overt persuasion. In *A Grammar of Motives* (1954), Burke described his theory of *dramatism* [sic, Burke's spelling], which considers human communication as a form of action best analyzed by using the vocabulary and concepts of the theater.[6] Whereas Aristotle, for instance, modeled rhetoric as a three-part construct of *speaker, speech, and audience*, Burke described rhetoric as a drama consisting of five interrelated components, the *act* (what was done), *agent* (protagonist, who acts), *agency* (through what means), *scene* (against what backdrop or setting – historical & cultural as well as immediate environment), *and purpose* (motive for action, the "why"). These five terms, named the *dramatistic pentad*, provide the necessary tools for understanding what motivates people to act in the dramas of everyday life.

The foundational importance of Burke's *dramatistic* theory of rhetoric rests on the search for motives: *What we say* that people are saying and doing, and *what we say* about why they are saying and doing it. In other words, Burke's rhetoric is all about discerning and attributing people's motives to action – the motives of the people when they speak or act, and the motives of people who respond, react, and critique them.

23

Traditional Greek rhetoric contrasts starkly with the contemporary Burkean dramatistic rhetoric. Burke departed from Aristotle's tripartite model of rhetoric that features an advocate, addressing a speech to an audience, in the context of one of the three selected kinds of public settings, for the purpose of persuading them. Burke, in contrast, took his starting point to be something very much like Aristotle's *The Poetic*, which is all about the structure of Greek tragedies. Compare the elements of Burke's rhetoric and the Aristotelian elements of tragedy:

Burke: Aristotle's *Poetic*

Act:*Mythos* or plot. Conflict; Protagonist *vs.* Antagonist, A climax leading to a New Order, or Restoration of Order.

Agent: *Ethos* or character. The Protagonist and Antagonist.

Agency:*Lexis* or speech; plus *Melos* or melody, music. How characters perform actions to accomplish their purposes.

Scene: *Opsis* or spectacle. Constraints on the action; background.

Purpose: *Dianoia* or thought, theme. What it means or accomplishes.

2. Some Implications of the Rhetorical Shift from Persuasion to Dramatism

A. Traditional Aristotelian rhetoric emphasizes rational argument, whereas Burke emphasizes drama— immediately we have expanded rhetoric *from* a mainly left-brained skill *to* a more balanced whole-brained process, with primacy given to the right brain.

B. Freudian and Jungian archetypes of character, action, etc., that make up unconscious human motives, become the stuff of rhetoric. Persuasion-as-identification becomes, among other things, the study of attitude and attitude change, amenable to the methods of social science, not just speculation about the supposed practical effects of a speech upon the audience. Rhetoric is a cooperative, interpersonal action between people, rather than an operation or manipulation performed by a speaker upon his audience.

C. Dramatism provides a new template for understanding, constructing, and evaluating the dynamics of public communication media. Twentieth Century rhetoric went through rapid growth and development with each new electronic technology—movies, radio, television, and now computers and the Internet. Think visual symbols and imagery.

D. Dramatism also taps into the unconscious, just as theater does, when it is reverse engineered back into the content of oral and written messages. Think of mythology, especially, the *monomyth* of the basic quest story, such as Joseph Campbell detailed in *Hero with a Thousand Faces*.[7] Later, in *The Rhetoric of Religion* (1961), Burke, an agnostic, detailed the Biblical *pollution-purification-redemption* schema as the fundamental *rhetorical* template for bridging divisions between people.[8] Evangelicals in particular should have no difficulty in grasping this principle; preaching, by and large, consists of telling and re-telling "the old, old story." The same redemptive dramatic construct works in secular rhetoric aimed at social action to solve problems. Most importantly, it cannot be overemphasized, according to Burke, the chief objective of rhetoric is to create identification between people, a *meeting of the minds,* rather than a speaker trying to *change* the audience's minds through argument and reasoning.

To recap: the Burkean pentad provides an intellectual apparatus for answering questions about *what we say* a person does or says, and *what we say* is the reason or the motive for it. Note also that the concepts and vocabulary of the dramatistic pentad are flexible, and can generate alternative insights into the rhetoric from the critical analyst's point of view.

3. Language is Symbolic Action. Language, according to Burke, is what makes us human. In *Language as Symbolic Action* (1966),[9] he composed a cryptic definition/poem that condensed his philosophy of man [generic *"he,"* consider the date]:

Man is the symbol-using animal, Inventor of the negative;
Separated from his natural condition by instruments of his own
    making;
Moved by the sense of order, And rotten with perfection.

Burke's poetic definition of rhetoric is embedded in his philosophy of man. He packs many powerful concepts into these few terms. Here are a few key interpretations. a). Language is *symbolic.* b). Symbols carry within themselves the *motives to act.* c). The *negative* incorporates the *moral, judgmental* aspect of symbolic actions. d). Everyday life is built around divisive motives between people; people are divided by the pressure of competing motives. e).

25

Divisions are disruptive of social order. Symbolic action is required to bridge divisions and restore order.

4. Burke's *Terministic Screen*

Burke's *terministic screen* is another one of those Kuhnian paradigm shifts in contemporary rhetorical theory. Considering language to be a *symbolic act* that *creates an identity,* words (and phrases, and pictures, and gestures, and all other symbols) have the potency to generate significant meanings. Words (and symbols) not only *reflect* reality; words *select and deflect* reality. (Like Freud, Burke liked puns and wordplay.)

According to Burke's *terministic screen,* words provide a way of interpreting the meaning of our reality in an ideological sense. Using critical shorthand, Burke talked about *devil terms* and *God terms.*

An example is when I once participated in a church mission visit to a refugee facility in Salzberg, Austria. Our little group was given specific instructions by the local host. We could distribute toys to the children, and try to be friendly and respectful. However, we were warned not to bring up our Christian affiliation by mentioning God or Jesus or our church, or even by wearing a cross, or in any other way to represent ourselves in a religious way. Most of the refugees were Muslims from Africa or the Middle East, where they had been taught that we Americans are *"The Great White Satan."* Our mission group intended primarily to be among their first impressions of human kindness from white Christians, without adding fresh pain by rubbing sectarian salt in their wounds. The first day we visited, four children showed up. The next day, forty children mobbed us and completely exhausted our stockpile of goodies for them, and some of their parents invited us to their rooms for tea.

God terms and *devil terms* need not be literal references to the deity. Screening terms may be any other words, phrases, or symbols that trigger an ideology in the public or social discourse. A single devil term or God term can be the catalyst to elicit entire clusters of ideas and visceral responses. Politically, think of the partisan rhetoric that divides America today. *Liberty, raise taxes, cut spending, Ground Zero mosque, health care, global warming (or is it global climate change?), jobs, bailouts,* even the names of *Obama, Nancy Pelosi,* and *Sarah Palin.* We could point to video loops on Fox News and MSNBC. Think of the unending repetition of Obama pastor Rev. Wright's shouted "God Damn America!" on *The O'Reilly Factor* during the 2008 campaign—or, on the other end of the spectrum, Keith Doberman's Countdown loop of the image of Rush Limbaugh bouncing up and down maniacally in a speech to conservatives. Those media symbols are *devil terms* that, all by themselves, call up clusters of ideological responses among viewers *who identify with those images negatively.*

I can produce many more actual published examples of it in the literature I have accumulated the past few months. Think, for instance, of the May 2010 hearings of the Texas Board of Education textbook adoption process, in which social conservatives on the Board put forward a resolution warning textbook publishers not to include "gross pro-Islamic, anti-Christian distortions" into world history textbooks that they view as being "rife with Muslim propaganda."[10] And, in a comment added to Michael Zimmerman's Huffington Post Sept 14, 2010 blog about that same Texas Board, someone signing in as "Mac Torquil" responded, "Is [sic] is sad but true that evangelicals, the same fools who gave us GW Bush, control Texas politics at the state AND local level. More and more, I am coming to agree with the HP poster who said, Religion is poison." –thereby *demonizing* the evangelical Board members as "fools," and their religion as "poison".[11]

The *terministic screen* is one of the instruments of the rhetoric of identification at the level of the individual symbol or word. Demonizing and valorizing—which make up the function of *devil* terms and *God* terms—serve to divide, and not to bridge our divisions. Unless, of course, we are all on the same page to begin with.

PRIMER ON WALTER FISHER'S NARRATIVE PARADIGM:
Further Developments in Twentieth Century Rhetoric

When I identify myself as a "narrative rhetorician," I mean it in the strongest possible way. Walter R. Fisher's *Human Communication As Narration: Toward a Philosophy of Reason, Value, and Action* (1987)[12] came at a crucial juncture in my professional and personal life. In 1986, when I was 47 years old, I was in the throes of a major midlife crisis. I had taken a new job requiring us to move from Houston to Richmond, Virginia. I had been coaching debate for about 25 years at that point and I was burnt out. I also underwent a deep spiritual renewal in my personal life that drove me out of my avowed intellectual skepticism, and back to my earlier Christian faith. To be honest, I was not sure I wanted to remain in higher education, teaching traditional rhetoric and coaching debate. I needed a total change in outlook.

The appearance of Fisher's *Human Communication as Narration* turned my entire worldview upside down. I took to narrative not only as an innovative, creative new academic theory of rhetoric. It explained my very life. For openers, I understood for the first time the roots of the transformative power of the gospel. For the gospel is nothing if not a story. So, Fisher enabled me to claim a respectable scholarly rationale for devoting all of that time and energy to studying the Bible, looking deeply into parables,[13] and Christian preaching – *as rhetoric,* as well as a strong personal interest. From there, I caught the insight that storying *as rhetoric* also legitimized my compulsive movie going. Rather than just entertainment, or mindless escape from my midlife problems, movies became as much rhetoric as the Gettysburg Address or M. L. King's "I Have A Dream." I came to define movies as *social texts* to read and criticize just as I would a speech. Since 2004, I have published a couple dozen movie critiques in *Christian Ethics Today.*[14]

I remained at the University of Richmond for eighteen years until I retired in 2004, and never coached debate again. From then on, everything I taught, and wrote about, revolved around narrative rhetoric, preaching as rhetoric, and the movies as rhetoric. Looking at my career overall, you would see that the first half might be considered *left brained, rational, traditional* rhetoric and argumentation, and the second half might be considered *right brained, narrative,* rhetoric of identification, or museums and monuments and historic re-enactments as cases of public memory; and especially, my newfound attraction to the study of preaching and the movies.

Fisher's Narration as a Human Communication Paradigm

The following table is paraphrased from Fisher's book, Ch. 3, "Narrative as a Human Communication Paradigm."[15] Fisher

28

compares and contrasts his narrative paradigm with the traditional rational worldview. Like Burke before him, Fisher begins with a philosophy of man (sic). Unlike Burke, Fisher progresses from the nature of man to the nature of rhetoric in the world, through a series of five sequential steps.

Here is the Fisher's summary of the Rational-World Paradigm, since Aristotle:

1. Humans are essentially rational beings.
2. The paradigmatic mode of human decision making and communication is argument—discourse that features clear-cut inferential and implicative structures.
3. The conduct of argument is ruled by the dictates of situations—legal, scientific, legislative, public, and so on.
4. Rationality is determined by subject-matter knowledge, argumentative ability, and skill in employing the rules of advocacy in given fields.
5. The world is a set of puzzles that can be solved through appropriate analysis and application of reason conceived as an argumentative construct.

In short [in the Rational-world paradigm], argument as product and process is the means of being human, the agency of all that humans can know and realize in achieving their **telos.** *The philosophical ground of the rational-world paradigm is epistemology. Its linguistic materials are self-evident propositions, demonstrations, and proofs – the verbal expressions of certain and probable knowing. [emph. Supplied.]*

The objective of this section of this essay is to lay out the Fisher's Narrative Paradigm by matching his list of five narrative presuppositions against those of the Rational-World Paradigm, printed immediately above. However, it is mandatory to underscore the clarity of Fisher's connection with Burke's mantra that *language is symbolic action.* There are many different philosophies of man, Fisher says, which may be expressed by certain "root metaphors." *Homo faber, Homo economicus, Homo politicus. . . ,* "psychological man," "ecclesiastical man*,*" *Homo sapiens,* and, of course, *"rational man"* [this latter philosophy being reflected in the Rational-World Paradigm.]

29

Fisher's philosophy of man takes the metaphor one step further, *Homo narrans,* or, man as storyteller. The crucial importance of viewing man first and foremost as a storytelling animal, rather than a toolmaker or calculator or saint, or even, as a philosopher, is this: in all of the other models of what makes us human, storying is always seen merely as a figure of speech, usually, a weak form of argument at best. But when man is regarded first as a storyteller, then *all the other metaphors become sub-plots.* **Narrative, thus, *subsumes all other modes of being human, including the rational*.**

Fisher's five presuppositions of the Narrative Paradigm are as follows. Keep in mind that each one is carefully paired with its counterpart in the list of presuppositions for the Rational-world paradigm.

1. Humans are essentially storytellers.
2. The paradigmatic mode of decision making and human communication is "good reasons," which vary in form among situations, genres, and media of communication.
3. The production and practice of good reasons are ruled by matters of history, biography, culture, and character.
4. Rationality is determined by the nature of persons as narrative beings—their inherent awareness of *narrative probability,* what constitutes a coherent story, and their constant habit of testing *narrative fidelity,* whether or not the stories they experience ring true with the stories they know to be true in their lives.
5. The world as we know it is a set of stories that must be chosen among in order for us to live life in a process of continual re-creation.

Theoretical Values Added by Fisher's Narrative Paradigm. Kenneth Burke was a prolific philosopher and literary criticism genius who wrote something on the equivalent of a book a year, most of which is *not* about rhetoric. Nevertheless, for the scholar, as we have seen, Burke is very rewarding. What can anyone possibly add to Burke?

Walter Fisher is one of the best interpreters of Burke that we have. Fisher not only organized and summarized Burke's thinking theorizing for more widespread understanding and application, he extended Burke's ideas. Here are some of the contributions Fisher's *magnum opus* made to rhetoric in the latter part of the Twentieth Century.

1. The five presuppositions of the Rational-World paradigm and the Narrative paradigm furnish a concise yet coherent picture of rhetoric as it exists within a macro scale, beginning with a conception of the nature of man, and proceeding from it to the nature of the world. If man is *thus*, then these are the rhetorical principles and operations that *this* sort of man necessarily must employ to navigate in that sort of man's world. It is no accident that, if man is considered to be a rational animal, then the world must also be considered a rational place; and therefore, the rhetorical process is also rational. Likewise, man, the storytelling animal, lives in a world of stories rather than a rational world; and the rhetorical process entailed by such a person in such a world must necessarily be narrative by nature.

2. In Narrative Presupposition 5, "The world is a set of stories to be chosen among," suggests that stories *as such* are neither true nor false, unlike the reasoning and argumentation necessitated in Rational-world Presupposition 1, "The world is a set of puzzles to be solved," which solutions must necessarily be *either true or false.* Story generates responses; a puzzle with only one correct answer generates arguments.

3. Fisher's provides *tests of narrative rationality* in Presupposition 4, *narrative probability* and *narrative fidelity.* This goes a long way towards reducing the logical objection to the ambiguity and indeterminacy of choosing between competing stories.

4. It follows that the Narrative paradigm legitimizes *standpoint rhetorics.* What are the wellsprings of stories in the world? Stories come from history, biography, character, and culture—all of which are manifestly subjective. How does one judge between the narratives, say, of Islam and Christianity? When I experienced my midlife *sturm und drang,* and felt the need for a spiritual transformation, I chose the story that rang most true to my history, biography, character, and culture, *i.e.,* I returned to the Protestant faith of my family in my youth. If I had been

born in the Middle East and reared as a Muslim, my choice of a new story to live by would have been predictably different.

5. Fisher's Narrative paradigm was developed in the 1980s in America, which suggests several other standpoint rhetorics of the day – feminism, Black rhetoric, Reagan Republicanism, and so on. Again, competing stories are available to be chosen among. Narrative rhetoric invites a response, not a debate. Rational-world rhetoric is the opposite—it invites either adherence, or a rebuttal—but not a personal anecdote.

6. Narrative Presupposition 2, positing "good reasons" as the basis for decision making instead of being limited to the more rigorous forms of logical argument, opens the door for truly cooperative and collaborative social action. Above all, the narrative paradigm is best suited to *moral* persuasion. The Narrative paradigm offers a powerful theory for the formation of belief communities grounded in mutual identification.

7. In no way does the Narrative Paradigm interfere or impede the Rational-world paradigm, although it subsumes it. Presupposition 2, about the paradigmatic mode of decision making and argument, in *both* the narrative and rational-world models, allow for situational and other differences. The important distinction to be made is, in the Narrative Paradigm, it is recognized that the scientific method is the best mode of decision making within the scientific context. That is so, because science is a discrete community that has its own commonly accepted worldview, standards, and procedures. On the other hand, in the Rational-world paradigm, the validity of storying, metaphor, and linguistic/symbolic rationalities, are rejected by its adherents in *all* situations, excepting stand-up comedy, movies & TV *for entertainment only,* more or less.

Creation/evolution is a textbook example. "Pure" scientists reject Biblical or cultural and mythical narratives outright, in or out of the lab (granting the exceptions that good scientists may also be believers in their personal lives); but those who are committed to the Narrative Paradigm up and down all five of its presuppositions are generally okay with science and technology, in the appropriate setting.

Language and Narrative: My Response to the Young Earth Creationists/Evolution Debate

To keep my approach simple, I represent Creationism by its most extreme version, Young Earth Creationism (YEC). That theological "take" is based on a literal reading of Genesis. YEC misreads the Biblical narrative, not as a collection of stories, but as dogmas, such that YEC is forced to reject all scientific evidence that undermines the position of the dogmas.

For YEC to deny all biological and physical evidence as a consequence of their unwavering belief in the literal meaning of Genesis 1, what I hear is not only that they believe God created the entire universe, out of nothing, within six calendar days. What I also hear is, their faith in God stands or falls on reading the Genesis text as the complete, infallible, verbatim words of God. To doubt or deny it is the same as denying that God exists. Framed in that way, YEC chooses God over not-God. When science denies Genesis, science must be mistaken, or perverse.

Language Use Differences at the Bottom of the YEC-Evolution Divide.

Let's frame the faith-science divide between YEC and evolution as a function of differences in language use. Later in this book, Robin Gibbons introduces in his essay, *For the Life of the World*,: "The problem lies in a number of areas not least the different types of language used to describe reality." I agree with Robin Gibbons. My aim is to narrow the distance between science and faith—as well as, between faith1 and faith2, and between science1 and science2. Cybelle Shattuck's essay also highlights the problem of different language usages to describe reality. Her essay exemplifies Burke's key theory, that language is symbolic action. She is a wise teacher to acknowledge, with her college students, "There are too many variables in play to make a one-size-fits-all language workable." Comparing YEC language with the scientific language of evolution, I want to underscore her point. Rhetoric does not pretend to provide a universal solution. In particular, narrative rhetoric celebrates diversity of different discourse communities. Her classroom is a place in which she proves every day that the world is a set of stories to be chosen among.

One more preliminary response to Shattuck's essay: The classroom is the context that informs her analysis. She describes the importance of college students learning to understand the many and various alternative stories surrounding the sanctity of nature. Students from the ages of 18-24 are still navigating through their post-adolescent identity crisis, equipping themselves to be fully independent, mature adults. This is not exclusively an intellectual process. It illustrates Kenneth Burke's theory of rhetoric as a process of identification, of internalization, rather than of persuasion. Shattuck's class is a life laboratory for students who are engaged in the necessary task of forging new adult identities for themselves, in the midst of myriad possibilities.

YEC is based on what James Fowler's stages of faith would label as a juvenile stage of belief in magic, and a religious identity anchored in submission to the authority of literalism, tradition, and conformity. The Fundamentalist insistence on adhering to a literal interpretation of Genesis is, in my opinion, not so much about the reliability of Scripture as it is about yielding to the absolute authority demanded by those who are interpreting it thusly.

Young Earth Creationism vs. Evolution: the Language Differences.

Several of our writers in this book—Denis Alexander, Darryl Falk, Rebecca Flietstra, among others - are Evangelicals who are fully engaged at the intersection between creation and evolution. (I will refer to them in more detail in the second main point in my essay, below.) I suggest that language, as a major piece of what it is to be human, can be usefully viewed as the core of that important creation/evolution intersection. Recall Burke's insight that language is symbolic action.

In prehistory, by definition, absent written language of any kind, (no repositories of public memory), let's adopt a common theory that language slowly evolved through a very lengthy series of progressive stages of magic first, then religion, then philosophy, finally science. I cannot make the case that this theory is accurate. Frankly, it is a controversial theory, especially about the primitive stages for which no evidence exists. Please grant for the sake of argument that language evolved through these stages in broad outline. The superstructure of the pre-existing oral language - the foundation, framing, rafters and joists—hardened into the archetypes (master metaphors), concepts, syntax and grammatical structures, abstractions, and the vocabulary that finally lifted humankind above the level of primitive magic and primitive religion. The warp and woof of mankind's first oral language are continually spun out of the golden threads of magic and the purple threads of religion, woven throughout the richer multi-hued mosaic of language as we know it in our lives.

Each one of evolution's infinite progressive changes and adaptations is based on what went before. Language shifts, likewise, are built on what went before. At first, human communication functions were simply gestures, rudimentary oral language, and visual symbols. What did language convey in the beginning? According to the theory of language evolution outlined above, it expressed a worldview of magic for up to 70,000 years; then later, of rationalizing magic into ritual and religion. Stages in the evolution of language out of primitive magic and religion were the products of evolutionary adaptations, until the invention of writing ushered in the emergence of more sophisticated philosophy and theology. The appearance of writing is based on the pre-existing oral language.

The language of science therefore inherently contains the language of magic and religion, as evidenced in the etymologies of scientific words in resources like the Oxford English Dictionary. Everything besides magic and religion evolved later. Written language - the ever-growing accumulation of tales, poems, laws, and histories - is necessarily embedded in the increasingly sophisticated modifications of those earlier symbols and ideas. If evolution has any meaning at all, how could it be otherwise?

The Language of Science and Religion.

The ultimate language of science is the latest outgrowth of that intrinsic lattice that marks the steady evolution of language. Scientific language is necessarily constructed and overlaid on earlier levels of language.

When did religion develop, as distinct from sheer magic? We do not know. For the sake of argument, let's pitch a commonly accepted benchmark, no doubt vastly oversimplified: writing was invented about 6,000 years ago. Whenever, it's a number more recent than the invention of primitive speech. The birth of religion as we think of it is associated and probably correlated more or less with the appearance of writing. The invention of writing gave a tremendous boost to the development of abstract reasoning. When language, and everything it signifies, is committed to writing, it can be carried over, stored, and built upon by communities. The key to the explosion of knowledge is that, without writing, we would have no idea of what we would be missing.

The three great Abrahamic religions, Judaism, Christianity, and Islam are all "religions of the Book." The Hebrew Bible by definition predates the Christian Era. The New Testament predates Christian creeds. To form Judaism, the Torah, Wisdom, Prophecies, and Histories were written [evolved] over a thousand year period. The Patriarchs, if they were indeed historic persons, predated Judaism. The Prophets predated the Second Temple Covenant. The Old and New Testaments combined add a few more centuries and several more entries into the anthology of sacred texts (books of the Bible).

Judaism is not the only religion to come about during that most fertile moment in the evolution of language. Bruce Feiler refers to some other developments in religion during the "Axial Age" of 800-200 B. C. E.: "Shintoism, Confucianism, Hinduism, Zoroastrianism, Judaism, and Platonism. . . .ideas of the individual and God, faith and reason, theocracy and democracy, church and state—were born in the centuries between Moses and Jesus."

What of the language of religion? One thing is certain: religion holds firmly to the vocabulary, metaphors, stories, and contexts originally given in ancient sacred texts. A major function of religion is to maintain and pass down the linguistic heritage that embodies religion from generation to generation forever. Verbatim texts of holy books are held as inviolable. Regardless of how different religions interpret their sacred texts, either hard line or soft, whether literally, allegorically or symbolically, the Torah, Tanakh, the Old Testament and the New Testament, and the Qur'an are all taken as the words of God, not to be gainsaid or tampered with.

One more thing is certain: religious language is not scientific. Religion is built up on the expression of wonder, awe of nature, narratives, laws and The Law, exhortations, proclamations, poems, prophecies, visions, proverbs, prayers, testimonials, dreams, miracle stories, myths, biographies of the saints and martyrs, and snatches of philosophy and theology. Religious language therefore is rich, figurative, evocative, inspirational, emotive, subjective, consoling, persuasive, poetic, personal, esoteric, and yes, mystical and mysterious—all of the qualities and features of language that science abhors.

Science admittedly uses models, paradigms, analogies, and metaphors all the time. Despite the rigorous editorial process through which scientific writing must pass in order to be published, still traces of magic and religion creeps through in language choices. For instance, I recall reading one essay in which the author made a point of inserting the term cosmos for creation, deliberately; yet later in the essay, the author lapsed into a reference to the animals as "creatures." [I regret that I have lost that reference.] But even if scientific language is completely scrubbed of all religious connotations on the surface, still the original meanings of terms and idioms remain. You can replace "kinds" with "species," and talk about animals as "specimens," all you want to. The fact remains, etymologically, the first uses of the term "species" referred to the bread and wine representing the body and blood of Christ in the Eucharist, and it also constitutes the root term in the notion of "special creation." This example is illustrative of many similar examples of how scientific terminology ultimately derives from religious language.

Unique Characteristics and Qualities of the Language of Science.

At last, our rough timeline of the evolution of language brings us to scientific language, which for the most part was invented only within the past few centuries. If our exalted Oxford colleagues in this book permit me, refer to The Cambridge Encyclopedia of the English Language.
Gutenberg invented the mechanical printing press in 1441. The ability to make many thousands of identical true copies in much less time than formerly required to reproduce a single copy of a book by hand changed everything. Within a century, the dissemination of written

knowledge went viral throughout Europe, and from there, to the far-flung colonies throughout the Western world and their closest neighbors.

The Royal Society for the Promotion of Natural Knowledge was founded in 1660. The Society's aim was to develop a plain, objective style, without flowery rhetoric and classical vocabulary, which would be more suitable to scientific expression. (p. 87.) English scientific and technical language expanded steadily throughout the Renaissance. The Nineteenth Century saw unprecedented explosive growth in the scientific domain(s), powered by the Industrial Revolution and unlimited frontiers of scientific exploration.

Each new discovery, such as electricity and evolution, introduced vast lexicons of new terms. New science journals were published. (p. 372). Typically, as science grew, many new terms appeared, including biology, taxonomy, ethnology, embryology, most of the nomenclature of chemical elements, many new precision measuring devices such as voltmeters, terms like watts and electrons, photosynthesis, spermatozoon, diatom, symbiosis, the Geologic Ages, and medical terminology such as laryngitis, cirrhosis, neuritis, and claustrophobia—and much, much more.

Besides the massive, complex lexicons of terms now understandable only to specialists, scientific language developed the following innovative distinctions that separate it from all other language usages. (p. 368)

1. It uses abbreviations, numerals, and special symbols.
2. The style is lexically quite dense: over 60 percent of the words in its prose are lexical (scientific words), less than 40 percent are grammatical (adverbs, adjectives, pronouns, connectors). Scientific terminology is careful to remain completely unambiguous.
3. Sentences average over 20 words in length.
4. The style is 99 percent impersonal. The use of "I" and "We" is frowned upon.
5. The passive voice is the norm. No active agents or guiding stories are imposed on physical processes. God is not an explanation; nor is the individual scientist. Scientific language is all about the science itself.

6. Noun phrases with complex phrases are usual (a typical example: transparent removable alignment grid for drawing external landmarks on the skin.)
7. The structure is compact; parenthetical phrases are used.
8. To repeat, just for emphasis: Narrative is assiduously edited out.

Exclusive reliance on these language conventions hampers the possibility of a real dialogue between science and faith. Science evolves as rapidly as possible. Religion tries not to change at all. Faith takes ambiguity and mystery to be the bedrock of spiritual (and aesthetic) enrichment. Science sees them as the enemy of progress. Unfortunately, the conventions of language usage in both faith and religion directly violate most of the boundaries of scientific language. For a scientist to try to adapt one's writing style to accommodate the subjectivity and ambiguities of personal narratives must feel like slumming, if not like a betrayal of one's identity as a true scientist in good standing.

But not vice versa. The scientists who are also devout believers in their religious traditions are motivated to bridge the divide, and to demonstrate that one can be a committed scientist, and also devoutly faithful. I find that as they do, they commonly sprinkle personal narratives and avowals of their personal faith, using religious language, throughout their scientific prose.

Conclusion

In the first part of this essay, I offered a primer of rhetorical theory, starting with Aristotle, and leaping over two millennia to touch on Kenneth Burke's theories of dramatism and the terministic screen, and Walter Fisher's narrative paradigm. In the second part, I have described, analyzed, and critiqued the rhetoric of Young Earth Creationism and evolution, regarding language itself. In the process, I have employed both rational and narrative rhetoric.

With these lines, I summarize what I have to offer, but I do not come to any conclusions. My hope is that I might have stimulated you to engage in further conversation. Ultimately, I fervently hope that we can leave this rhetorical workshop behind, and use it to advance our dialogue of the sanctity of nature. Towards that end, I want to finish this essay with a famous paragraph from Kenneth Burke known as "The Unending Conversation."

Imagine that you enter a parlor. You come late. When you arrive, others have long preceded you, and they are engaged in a heated discussion, a discussion too heated for them to pause and tell you exactly what it is about. In fact, the discussion had already begun long before any of them got there, so that no one present is qualified to retrace for you all the steps that had gone before. You listen for a while, until you decide that you have caught the tenor of the argument; then you put in your oar. Someone answers; you answer him; another comes to your defense; another aligns himself against you, to either the embarrassment or gratification of your opponent, depending upon the quality of your ally's assistance. However, the discussion is interminable. The hour grows late, you must depart. And you do depart, with the discussion still vigorously in progress.

From the Editor....

In the previous essay, David Thomas demonstrated the importance of recognizing rhetoric, especially in debated issues. As we proceed in our attempt to discover the sanctity of nature and our human place within it, we find it necessary to first cross a difficult and problematic bridge. Where did nature come from? How did it get here? How did humans arrive? Our ultimate mission can so easily be derailed by these questions. To be sure, many of us hold very clear and passionate beliefs. Yet, it is humbling to recognize that our beliefs are developed within the scope of rhetoric and narrative, and that we cannot think outside of language. Language is such an important part of any discussion, and it is certainly an important part of a discussion on the issue of nature and our human place in it.

It would be a lofty goal to think that we could find some commonly accepted language around which all of us could unite. This goal would be as unrealistic as it would be lofty. Perhaps such a goal should not be pursued. Rather, is it possible to allow people to utilize the language and narratives with which they most relate in order to find ourselves united at a common place and around a common cause? This is the question at the heart of our next essay.

Our next writer is Cybelle Shattuck, who has a graduate degree in Religious Studies from the University of California, Santa Barbara. She has taught for Western Michigan University and Kalamazoo College, where her courses focused on Religion and Nature, Hinduism, Women in Asian Religions, Judaism, and Religion in America. In this essay, Shattuck reflects on the challenge of discussing the issue of the sanctity of nature with college students who have been unable to reconcile their faith language with their scientific language. Recognizing the limits of language, Shattuck looks for other possible avenues of finding unity.

Chapter 3
Expressing the Sanctity of Nature

Language for experience, inspiration, and building bridges

by Cybelle Shattuck

Teaching a course on Religion and Nature for a Midwestern liberal arts college provides a snapshot of the struggle to reconcile science with faith. The students represent different denominations of Christianity and Judaism, as well as the "spiritual but not religious" community, are mostly middle-class, and are intellectually curious. Many of them are preparing for careers in medicine and most have a strong desire to do work that helps other people. They are drawn to the class by their deep connections to nature, which they experience as sacred. But they are also uncertain about how to describe that experience because the theological language they absorbed from their religious communities while growing up does not align with their scientific knowledge. They cannot reconcile a biblical deity who micromanages creation and intervenes in history with the theory of evolution, yet they do not have an alternative vocabulary for describing their sense that divinity exists in some way.

These students were not merely intrigued by readings that offered them new vocabularies, they were grateful to discover that they were not alone in their frustrations or in their intuitions. Recognizing the depth of their desire for a new language highlights how important this issue is to the perceived conflict between religion and science in the US. It may also explain why this theme emerges in several of the essays in this book. This essay explores some of the new languages that are emerging to address this need, but it also asks if revised language alone is sufficient to reconcile the many tensions in current interfaith and faith-science conversations. Perhaps a shift in the focus of the conversation can offer another way forward.

New Vocabularies

The new vocabularies students encountered in the class readings of my course took diverse approaches to describing the sanctity of nature. Some had strong Christian themes while others moved beyond biblical traditions to focus on Asian philosophies or even the use of scientific knowledge itself as a vehicle for spirituality. No one language resonated with all students, but there were discernable patterns among the texts that students preferred. These generally presented the sacred as an impersonal Energy rather than a personified Lord, thereby eliminating the problems associated with a deity acting on the world in violation of natural laws. The more popular texts also included ideas drawn from modern science and economics, while placing a strong value on environmental ethics.

The most Christian of the texts in the course was Sallie McFague's *Life Abundant*. This book delighted some of the students who had given up on Christianity because the only interpretation of God they had previously encountered assumed a fairly literal reading of the Bible in which God is personified and active in history. There are, of course, multitudes of theologians who provide sophisticated interpretations of divinity in more abstract terms such as Godhead, etc, but these are not part of the elementary Sunday school curricula so they have not been absorbed into public American God-language. Consequently, McFague's description of an abstract Divinity that underlies the universe and is manifest in the entire physical world was a revelatory reading of the Bible for these young people. In her version of Christianity, there is no separation between a sacred God and profane world.

To me God is reality and the source of our reality. While God and the world—God's reality and ours—are not identical, they are ontologically related. That is,   the world's reality derives from God, but just as important, the world is God's   beloved which is joined to God: the world is God's body.[16]

In McFague's theology, the divine incarnation is not limited to the historical Jesus, it refers to the embodiment of God in the physical world such that "each creature is a microcosm of the divine incarnation; each of us is made in the image of God."[17]

This idea allowed students to reconcile science with religion. It matched up with physics classes that taught them to see the objects

43

in the universe as energy organized in specific ways that look different on the surface, but are really part of one interconnected field. It could be reconciled with biology lessons that described all life forms as branches of one great tree—all with a common origin and some measure of shared qualities.

This emphasis on the connections between humans and other creatures on the planet then becomes the basis from which McFague critiques consumer-based economics. She argues that a truly moral economic model should consider how practices affect all inhabitants of the oikos (household)[18] , which requires factoring in resource extraction and long-term impacts when calculating cost benefits. She takes up the theme of Christians being called to aid the poor and oppressed and applies it to all life on the earth, suggesting that an economy that damages biodiversity is causing oppression and poverty to members of God's household. This practical application, linking ecosystem-based interconnections with morality, struck a chord with students. It brought together science and faith to create an ethical framework for challenging consumerism and promoting service, thereby validating their desires to seek meaningful work.

Students were clearly attracted to vocabularies that stressed interconnections and matrix thinking such as Buddhism and, to a lesser extent, Wicca. Buddhism appealed because it described an evolutionary universe, in which everything is characterized by impermanence. Things (including people) come into existence as aggregations of components and exist as coherent beings until the parts disaggregate and recombine to form new beings in a continuous cycle of rebirth. This model worked for the biology students, who likened it to a cellular being. To them, it made sense to describe a person as a collection of parts that fulfill certain functions and that the future form of the person would be affected by the actions taken in the current state. So, for example, if the person ate sensibly and exercised, then the cells would form healthy new cells to replace themselves in a kind of cellular cycle of rebirth. But if the person smoked or was exposed to pollutants, then the cells might evolve into cancerous forms.

The most attractive aspect of the Buddhist worldview was that it could connect reincarnation with evolution while also adding a way to reclaim the idea that there is a special purpose to human life. This helped counter the most difficult implication of evolution, the idea that human beings evolved randomly with no special purpose beyond reproduction and survival, so our lives are not ultimately meaningful. Buddhism emphasizes that humans are merely part of the interconnected universe, with the same basic characteristics as everything else that exists, but describes them as special because they can attain wisdom and make choices about their behavior. The ideal path for humans is to seek knowledge about themselves and the true nature of the universe. When people recognize they are connected to all the other beings making up the fabric of the cosmos, their lives are transformed and their behavior is shaped by compassion for others. This ethic provided a way for the students to connect scientific study with their desire to do work that would help others, especially when they encountered Engaged Buddhism through the writings of Joanna Macy, who stresses that realizing one's interconnectedness with all beings can be an inspiration for environmental action.

A similar matrix emphasis appears in Wiccan worldviews. Some students were struck by the idea of the world as a manifestation of energies that could be accessed through ritual—not unlike the way physicists produce energy through specific manipulations of particles. They interpreted stories of supernatural beings as ways to allegorize abstract energy, and this gave them new insight into possibilities for interpreting other scriptures. They were also intrigued by the emphasis on observation of nature as a means to learn about earth's energies and how to live in harmony with a greater natural order.

A number of students found their most relevant narrative for harmonizing natural order and human life through biologist Ursula Goodenough, who offers a vocabulary she calls "Religious Naturalism" as a non-theistic alternative to traditional religions. She describes the goal of her book this way: *A cosmology works as a religious cosmology only if it resonates, only if it makes the listener feel religious. To be sure the beauty of Nature—sunsets, woodlands, fireflies—has elicited religious emotions throughout the ages. We are moved to awe and wonder at the grandeur, the poetry, the richness of natural beauty; it fills us with joy and thanksgiving. Our response to the workings of Nature, on the other hand, is decidedly less positive. The scientific version of how things are, and how they came to be, is much more likely, at first encounter, to elicit alienation, anomie, and*

45

nihilism, responses that offer little promise for motivating our allegiance or moral orientation. It is therefore the goal of this book to present an accessible account of our scientific understanding of Nature and then suggest ways that this account can call forth appealing and abiding religious responses—an approach that can be called religious naturalism.[19]

She then celebrates the awe-inspiring state of mystery that comes with knowledge of quantum physics and the chemistry of primordial life forms, the special sense of personal identity that arises from knowledge of how humans evolved, and even how the processes of cell biology can make life and death meaningful. She also explains why she cannot believe in a supreme deity, even though she would like to because she knows how much comfort the idea brings to many of her friends. Her honesty seems to free some students from a sense of guilt about their inability to reconcile science with the faith of their childhoods and gives them greater freedom to start developing new ways to articulate their own emerging worldviews.

Flexible Stories

Here then is a collection of new vocabularies that students found useful for expressing the sanctity of nature. One reason these theologies harmonize with scientific knowledge is quite simple—they do not have creation stories that are at odds with physics. McFague is an extreme example of Christian theologians who reject belief in literal interpretations of scripture. Buddhists do not have a creation story to contradict since Lord Buddha refused to explain how the universe came into existence on the grounds that this knowledge would not contribute towards liberation from the cycle of rebirth. Wiccans take a position similar to the Buddhists, focusing on a description of the world as it is rather than worrying about how it began, but they also are comfortable interpreting sacred narratives as allegories and metaphors rather than literal accounts. For Goodenough, physics is part of the sacred narrative, so there can be no conflict.

But this is not quite like Thomas Berry's vision of a New Story or Earth Story, a sacred narrative recounting the history of the planet that he hoped would provide a common narrative and bring us all into accord. No single vocabulary worked for all the students. It seems probable that this will always be the case because there are too many variables in play to make a one-size-fits-all language workable. These students had already moved away from the creation stories they learned in childhood so most of them were starting from a scientific frame of reference and looking for ways to bring in spirituality. People who begin with a specific religious framework and try to bring in science will evolve different sets of vocabularies.

This inevitable diversity does not, however, mean that there is no hope for a new, more fruitful conversation between science and religion. There was another common characteristic running through the texts that the students liked best—they all advocated environmental actions. And this may be the key to a common language that can cross philosophical divides and even build a bridge between religion and science.

Ethics and Environmentalism

What would happen if we changed the focus of the conversation from how beliefs about nature differ to why people care about nature? If a person of faith asked a scientist, "Why are you so worked up about global warming? Why does it matter to you whether or not I believe in it?" What kind of answer might these questions elicit? Would the scientist say something like, "I have kids and I want them to breathe clean air and eat healthy food so I care about reducing greenhouse gases and pollutant. I want my kids to live without fear of floods, droughts, food shortages, and energy crises. I want them to experience the forests and lakes that I enjoyed, to see eagles and robins, to pick blueberries and apples. Furthermore, I do not want them to live in a world where people are dying from malaria or starvation and they have to feel sorrow and helplessness in the face of widespread suffering."?

How might the religious person respond to this statement? Would she say, "I have kids too. And I want them to live in a moral world, where people care about their neighbors, and live by the Golden Rule. I believe that this planet is God's creation and we are going to be held accountable for what happens to it and the creatures within it. I share your concern for the poor and the suffering, and think it is important to live without causing harm to others."?

In this hypothetical dialogue, the scientist and the believer have shifted from arguing over who has the correct beliefs, to a discussion about their goals, values and ethics—and on these subjects, they can find some common ground.

There are different types of interfaith dialogue—and the conversation between science and religion is, at heart, about two different worldviews trying to communicate, so it fits these models too. Sadly, the most common style of dialogue is an effort to explain why one's own worldview is correct so that the other person is convinced of the "truth" and comes around to this same perspective. It is difficult to avoid doing this; we all want to prove we are right. But an alternative is to accept that worldviews are different and to assume that people have good reasons for believing in their own traditions. If discussants begin from this premise of mutual respect, then they are freed from the necessity of converting each other and can direct the conversation toward other ends. The most fruitful dialogues seem to arise when people focus on ethics. Apparently, even when people have quite different conceptions of what they should believe, they often have very similar ideas about how they should behave. Because of this, there are numerous examples of interfaith efforts to relieve poverty, feed the hungry, abolish slavery and, most recently, promote environmental justice.

The shift from an emphasis on who has the correct beliefs to an exploration of whether there can be shared actions requires recognition of common problems. Further, it requires information about what actions might best solve those problems. This is where science becomes an especially significant voice in the conversation. Scientists have amassed a wealth of information about how the world functions and how what happens in the environment affects human lives. Studies showing that pesticides affect fetal development provide a basis for regulating chemicals. Quantifying the links between air quality and respiratory diseases like asthma helps convince people that regulatory policies must be implemented. Modeling and experiments show that changing the albedo of roofing can reduce the urban heat index, thereby protecting people from heat-related illness, while also reducing the need for air conditioning, which in turn reduces use of energy that may come from coal, which had to be mined using processes that destroy ecosystems, pollute water supplies and, when burned, will add pollutants to the air that contribute to illnesses and climate change. We know about this chain of effects because of science and, with this knowledge, we can make better-informed decisions about how to change things. Consider how many different social concerns can be connected to this chain, all of which might provide incentives for people to take action. It is not necessary to have identical worldviews, or even for everyone to accept climate change, to agree that reducing coal burning is important.

Threatened Identities

Even if members of different faith and science communities discover that they have shared values and wish to engage in shared actions, interfaith work is not simple. Language can provoke further problems if there are miscommunications, and many people experience a genuine fear that finding common cause may threaten personal identities. The latter concern can be especially prominent for religious conservatives when the shared values and actions concern environmentalism, which comes burdened by old associations with a "liberal secular" label. Moving beyond old patterns of thought and the accompanying identity boundaries takes conscious effort. It may require that people re-evaluate the language within a tradition to align it with new goals and carefully consider how to incorporate new vocabularies for communicating with people of science and other faiths.

Rabbi Cohen-Kiener, who has done interfaith environmental work for many years, notes that when people from different communities who are trying to build interfaith alliances begin to expand their language in an effort to be more inclusive, they can start to feel that they are not being fully loyal to their own traditions. This is exacerbated if members of their own congregations accuse them of becoming "pagans" or earth worshipers. Such pejorative comments indicate a discomfort that may come either from a sense that environmentalism is not compatible with conservative religion or from concerns about the implications of accepting other faiths as valid. Cohen-Kiener describes this tension between holding to a distinct identity as a follower of a specific religion and wanting to work across traditions because restoring the earth requires cooperation.

As a rabbi, I am called by the Jewish tradition to perpetuate a particular religious tradition. Creation care pulls me in the opposite direction. The mandate to stewardship calls me to work with many others. Those others may disagree with me about a just solution in Israel/Palestine. Those others may disagree about the divinity of Jesus or the journey of the soul after death. They might disagree with each other about these and many other issues. But we cannot get the job done in our base camps. We must work together.[20]

She then goes on to recount a conversation with an evangelical in which both discovered that they were concerned about environmental justice and both agreed that economic institutions ought not to be treated as if they were somehow outside the purview of morality.[21] During this exchange, she learned about some theological teachings that the evangelical used to frame his sense of environmentalism and she shared some of the doctrines from her own tradition that inspired her efforts. When the conversation ended, these two people of faith knew that they could work together toward a common goal, yet each was secure in his, or her, own religious identity. Having a strong foundation in their own traditions and deep knowledge of how environmentalism fit into their faiths made it easier for them to interact as mutually respectful equals.[22]

Language as a Foundation for Action

Perhaps one reason some people of faith are so uncomfortable with the ideas of sacred nature, environmentalism, and science derives from another side to the language issue. As my students demonstrated, language for expressing the sanctity of nature has not been part of the regular doctrinal vocabulary for most Americans. Congregations that wish to make environmentalism a core value have to develop ways to connect this ethic with their own doctrines, just as my college students needed to find new ways to express their beliefs. Once this language is in place, then it becomes possible for people to take action and to have the strength of faith as a support when the path is less than straight.

Most of the rapidly expanding literature on eco-theology in the US attempts to reconcile environmentalism with doctrine. The earliest writings tended to focus almost entirely on beliefs, refuting Lynn White's allegation that Christianity was inherently anti-ecological by preferring green interpretations of scripture. Newer texts place greater stress on actions that people should be taking because of these green theologies. Programs range from Creation Care among evangelicals to social justice among Catholics and an Eco-kosher movement within Jewish communities. Running through most of these is a theology of stewardship rooted in scripture, doctrine, and tradition that is now being connected with conservation and environmental justice. Faith provides the platform for taking action.

The *E Pluribus Unum Project* provides a good example of how this can work. This interfaith program exploring religion, social justice, and the common good was launched in 1997, by Rabbi Sidney Schwartz. He felt that religion could be a source of great strength to those working for social change. As an educator, he believed that study of Jewish texts and values could be used to inspire greater commitment to issues of social justice and political activism. He created the EPU project to explore ways to motivate a diverse group of young people to pursue social justice as a common good.

The program had three objectives. First, it divided the students into study groups to learn what resources their respective religious traditions offered for addressing modern social issues. Rabbi Schwartz found that young people had little awareness of the social application of their own religions. At the beginning of the program, fewer than two-thirds of the participants could name three teachings that spoke to a social issue, but 95% could answer that question by the end. This grounding in their own traditions was designed to motivate the students to take actions, but Schwartz also hoped it would make it easier for the students to interact confidently with people from other faiths so they could find allies to help them in their work.

This led to the second goal of the program, to allow students to explore similarities and differences among their separate faith traditions and discover areas of common interest. Less than a third of the young people could name three social teachings from other faiths prior to taking part in the program, but more than 80% could do so afterwards. The third program objective was to teach participants how to become effective advocates for social and political change that would be informed by religious teachings. The theme of social justice enabled participants to look beyond their separate faiths and explore

51

the shared ethics that could form a basis for working together toward a common good.

A follow up study six months after the program suggests that this sequence of first establishing a solid foundation of study in one's own tradition, then investigating other religions, did facilitate interfaith cooperation. It found that "those most grounded in their own tradition were able to create the strongest relationships with people of other faith traditions in the pursuit of some social justice cause."[23] Surveys of alumni conducted two years after the program found that there had been a shift in students' religious self-perception. They "now framed their statements of belief and commitment from a much deeper place within themselves, coming to feel that they more fully owned their convictions of faith."[24] At the same time, they were much more confident about interfaith work. As one student said,

[W]e all discovered that the idea of working for the community and (for) the common good and going out and making a difference is something that is common in all our religions. . . . Before last summer I'd have been against working collaboratively (with people of other faiths). I would have only worked to support (my religion's) organization[25].

Another student also commented on how valuable it was to learn about shared values that could motivate shared actions.

Something I got out of EPU last year is looking at the religious texts from our various backgrounds and what our faith in general has to say about the quest for social justice. When you realize that they are so similar . . . you realize how much bigger a group of people from different faith backgrounds you have to work with for the common good. It is a very empowering thing.[26]

The E Pluribus Unum Project is an inspiring example. The participants developed vocabularies based on their own faith traditions that strengthened their commitment to action for social good. Once they had this foundation, they were able to partner with people from other faiths who shared the same goal, without feeling threatened by the use of other languages. Nor did they feel that everyone had to have all of the same beliefs in order to work together, they only needed to agree on social justice as a common value. Religious and science communities that share a deep concern for the environment ought to be able to follow a similar path toward collaborative work.

Promise and Limits

Religious communities are a tremendous resource for social change. They speak to a society's values and our environmental crisis is a crisis of values, so it is essential that religious voices be part of the conversation about how we can change our trajectory. Concerns for social and economic justice provide common goals that bring many faiths, and science, together. This is reminiscent of earlier abolition and civil rights movements, in which ethical values helped mobilize protest across religions, races, and classes to clear a path for change.

It is not necessary for everyone who shares these environmental goals to also share a single theology, story, or language for articulating a unified worldview, either scientific or religious. People are deeply attached to specific cosmological interpretations and these cannot be changed without setting off a cascade of reinterpretations that will affect beliefs about the nature of God, the human soul, morality, and salvation. Consequently, although we can describe the characteristics of theologies that may be most compatible with science, a dialogue between science and theology aimed at changing each other's beliefs is unlikely to be fruitful. Of more value is to seek within our own traditions for resources to validate convictions that nature is sacred and use these to provide support for our efforts to heal the damage around us. In this way, both religious convictions and scientific knowledge can motivate environmental actions.

Anchored by confidence that environmental actions are legitimated by one's own religious teachings, it is possible to use shared goals to build a bridge across the chasm between religions and between science and religions. A strong anchoring foundation assures that crossing the ridge will not threaten one's identity. Bolstered by confidence in our own beliefs and the ethics that go with them, we can reach out to explore the ethical codes of other worldviews and form partnerships where we find shared goals and values. There are many ways to express the sanctity of nature, but the effects of pollution and climate change are a shared experience, the causes derive from shared patterns and institutions, and the solutions require shared actions. Seeking those solutions may be the goal that inspires collaborative action among people of diverse faiths and science who are willing to step out onto the bridge.

Bibliography

Braxton-Little, Jane. "God and Greens." Forest Magazine: March/April 2001.

Cohen-Kiener, Andrea. Claiming Earth as Common Ground: The Ecological Crisis through the Lens of Faith. Vermont: Skylight Path Publishing, 2009.

Goodenough, Ursula. The Sacred Depths of Nature. New York: Oxford University Press, 1998

McFague, Sallie. Life Abundant: Rethinking Theology and Economy for a Planet in Peril. Minneapolis: Fortress Press, 2001.

Schwarz, Sidney. "Exploring Religion, Social Justice and the Common Good." The Reconstructionist: Fall 2000

From the Editor....

In the previous essay, Cybelle Shattuck suggested that rather than seeking commonly accepted narratives and theologies, people of diversity could unite around goals, and particularly the goal of recognizing the sanctity of nature and our human responsibility for its care. Indeed, we have found that to be true in the project that led to this book. As will be evident as we proceed, we have gathered together a group of thinkers who have very diverse religious identities. We may not find agreement on many of the details of our beliefs, but we did find real unity, and true community, around the theme of the sanctity of nature.

We may be willing to respectfully allow diverse theologies to exist side by side, mainly because as individuals, we don't necessarily have to integrate them within ourselves. I am perfectly willing to discuss common goals of creation care with a Muslim scholar such as Basil Mustafa, without having to incorporate any of the Islamic belief system into my own.

There is, however, one issue that almost demands some kind of integration. It is the relationship between faith and science. Although I can live my whole life without having to adopt a practice of Islam, I cannot live one day without adopting a result or practice of science. For all of us, faith and science must work out an integrated place within us. For the purposes of discovering the sanctity of nature and our human place within it, this integration is also necessary.

It seems that those among us who have the most difficult time working out this integration is our Evangelical Christian friends. Evangelical Christians have a long and rich history of an allegiance to a more literal reading on the Bible, including its account of creation. Since science has presented a case for evolution that calls the literal reading of the Bible into question, many evangelicals have seen religion and science as enemies.

If you consider yourself an evangelical Christian, we want you, and need you, in this journey. Many of our contributing authors are evangelicals. In many ways, I consider myself one. Without a doubt, that was the environment of my upbringing.

Can evangelicals come to peace with science? That is the subject of our next essay, written by Dr. Darrel Falk. Dr. Falk is the

President of The BioLogos Forum, an organization founded by Dr. Francis Collins, who is now the head of the NIH. Falk is also a professor of biology at Point Loma Nazarene University in San Diego. Beginning with his own personal story, Falk illustrates how one can embrace the world of science without diminishing an evangelical heart.

Chapter 4
How Can Evangelicals Come to Peace with Science?

by Dr. Darrel Falk

Formative Years

My interest in helping evangelicals come to peace with science is deeply personal. I grew up in a home that was firmly committed to evangelical Christianity. Among my very first memories (if not the first) is of an encounter with what we evangelicals refer to as Satan— the Devil. I was four years old and money was tight for my family. It was peach season in British Columbia and I remember being in a grocery store asking my mother to buy us some peaches. The fact is that we could not afford fresh peaches, so she said "No." At that point, Satan, I came to believe, moved onto my shoulder and whispered in my ear: "Take the peach. Take the peach." I listened to Satan, not my mother and I slipped the peach into my pocket and left the store. When I got home, not being particularly astute, I listened to Satan a second time. I pulled the peach out of my pocket sitting on the front stairs of my home and started to eat it. "Where did you get that peach?" my nine year old sister asked. Satan whispered in my ear: "Tell her that your mother bought it for you." So I did. I listened to Satan. I lied. These are horrific moments for an evangelical; I had given in to Satan, the same Satan who had spoken through a snake to Eve, and I had given in to the pressures of sin, just like Eve did. I ate of the fruit. I was four years old, and it is my first memory.

However, I also remember what happened next. My sister told my mother about the fact that I had listened to the Devil and the next thing I remember (indeed I remember nothing else), my father had come home from work, and I was crying as I, a four year old boy, knelt by one of our dining room chairs and told Jesus how sorry I was for listening to Satan rather than to him. With tears running down my face, I asked him to come back into my heart again…and he did!

I never saw Jesus; I never saw Satan, but there was this other world in which I was deeply steeped. It was a spiritual world, a world that couldn't be touched but was even more real than the world that could be touched. I loved that world. I loved listening to Jesus. To the best of my knowledge I never again told a childhood lie, and I certainly never stole a peach again. I wanted to live for this invisible Jesus and I did.

Hebrews 12:22-29 states: *You have not come to a mountain that can be touched, and that is burning with fire...you have come to Mount Zion, the heavenly Jerusalem, the city of the living God. You have come to thousands upon thousands of angels in joyful assembly, to the church of the first born to the spirits of righteous men made perfect...therefore since we are inheriting a kingdom that cannot be shaken, let us then be thankful and so worship God with reverence and awe for our God is a consuming fire.*

My most indelible memories include those of my 85 year old grandmother who lived in our home, but whose heart had long since left this world as she recited over and over again, "My heavenly home is bright and fair...O how I long to be there."[27]

I grew up tapping my feet and merrily singing: "This world is not my home, I'm just passing through. My treasures are laid up somewhere beyond the blue. The angels beckon me from heaven's open door and I can't feel at home in this world anymore."[28]

I smile about all of this now, but as I grew older and entered my teen years, I suffered considerable cognitive dissonance. I could not put the two worlds together. I found I was interested in science, and that I was pretty good at it. Whenever I moved into my science mode, I struggled mightily—especially if it was biology. I made it through those years as a Christian but only by refusing to think about biology, and focusing only on physics, chemistry, and math.

Eventually, however, the magnet drawing me into biology was just too strong. The inner world of the cell was too fascinating—too beautiful actually—for me to stay away from. I got sucked in by genes, ribosomes, and messenger RNA. . Along with that came at least some cognitive dissonance, and the evangelical faith of my youth disappeared. Jesus was gone, prayer was a thing of the past, and life, so I thought, would proceed with a focus only on the natural. My childish world of the supernatural disappeared and along with that went my cognitive dissonance.

The Evangelical Disconnect

The story of my return to evangelicalism is a long story for another time, but the point I wish to make is that since most evangelicals are not scientifically inclined, they simply don't come to learn how strong scientific data really is. They are able to live out their lives believing that:

.   • Since science can't detect God or the presence of God, and even seems to claim the absence of God, it can't be trusted in relation to any matter where it seems to conflict with evangelical dogma. Doctrine, evangelicals believe, comes directly from God through a plain reading of the Bible. Since science doesn't accept this it can't be trusted.
   • Since science can't measure God's response to prayer and even seems to claim that there is evidence that God does not respond to prayer, science's statements about the origin of the natural world can't be trusted. The world, evangelicals believe, originated for the express purpose of two-way communication between God and humankind. If science doesn't accept this as the basis for origins, what more can it add to the conversation?
   • Since science has no instruments to examine the Holy Spirit and even seems to say there is no Spirit in this world, it can't be trusted in any matter where it seems to conflict with the Word of God which was divinely inspired by that very Spirit
   • Since science leaves evangelicals feeling cold on matters related to purpose and meaning in life and since the Bible is so rich with purpose-filled directions for humankind, it can't be trusted to provide any helpful information about humankind's place in nature.
   • Since science seems to imply that there is no eternal life that transcends the present, evangelicals don't trust any statements it makes about the distant past either. So when it says the universe is 14 billion years old or that the chimpanzee and

59

humans had a common ancestor about 6 million years ago, the conversation ceases. Science can't be trusted to "tell time" correctly.

- Since science cannot study heaven and even seems to imply there is no such thing, why should scientific judgment be trusted when it speaks of the earth as though there is nothing more? Scientists speak about human responsibility in caring for nature in the present as though there is nothing else. Heaven is eternal and evangelicals see things only through the lens of eternity—something which science does not do.

I have spent the past 26 years teaching science to evangelical young people. Many have grown up in Christian schools which exist in no small part to help them see that science has been badly corrupted by principalities and powers that are not of this world. Even if they haven't grown up in Christian schools, almost all have grown up in evangelical churches. Although they come to university to learn science, many if not most have come to science not out of a motivation to study nature for its own sake, but rather to study it so that they can prepare for lives of serving God and humankind through medicine and other similar endeavors. Evangelicals are not, in general, fond of science for its own sake.

I want to emphasize that I remain an evangelical. Not only is my experience with sin and forgiveness as a four year old boy my first recollection, I still consider it my most defining moment. My understanding of the Christian life has grown considerably since my childhood and teen years, but I am a thoroughly committed evangelical. I hold to its tenets and my life remains firmly grounded in them. However, I am also a person who has spent my career in science. I trust science as being a God-given tool to understand this universe at a level that amazes me more with each passing year. Science is a gift of God, not a corrupt tool commandeered by Satanic forces. It tells us about God's world and does so in a manner that complements God's Word. Evangelicals can learn much about God and the activity of God by studying nature through scientific tools. The current closing of the evangelical mind to the world of science needs to transition from a world behind closed doors and shuttered windows to one which opens onto a huge porch with spacious vistas and deep blue skies. Out on that porch, gentle Spirit-inspired breezes need to draw Christians into worship as the world of science reveals the glory of God.

Building the Bridge

The current situation is untenable for those of us who love both science and evangelicalism. Since we hold that the evangelical approach to the life of faith is well along on the road to truth, and since we hold that science is an extremely effective way of understanding how God works in nature, then it is important that they not be at odds. The question, however, is how to construct the bridge that draws them together.

The task is not going to be easy. For example, the leading spokesperson for America's largest protestant denomination, Albert Mohler (President, Southern Baptist Theological Seminary) recently gave a speech in which he asks the question , in essence: Why Does the Universe Look so Old when it really isn't?[29] He proceeds to elaborate on the view that all of Christian theology falls apart if the earth is old. So it must be young. We will someday understand why science seems to point to an old earth when it is really young, but for now, he believes, we must accept that our knowledge is incomplete. These words are not the words of a crank on the fringes. Time Magazine has declared Al Mohler to be the "reigning intellectual of the evangelical movement."[30] Mohler has masterfully engineered the shift of the Southern Baptists' flagship seminary and in so doing has carried a huge denomination along with him. His successful influence on evangelical Christianity is the cover story in the magazine *Christianity Today* in the month of this essay's writing.[31]

Hugh Ross, of the highly influential Old-Earth-No-Macroevolution organization, *Reasons to Believe,* thinks that the whole purpose of the first 3.8 billion years of life's history is to prepare a place that is ideally equipped for human civilization. Beginning especially with the Cambrian explosion, according to his book, More Than a Theory , God engaged in a flurry of activity: *...the Creator worked efficiently to rapidly prepare a home for humanity. A huge array of highly diverse, complex plants and animals living in optimized ecological relationship and densely packing Earth for a little more than a half billion years perfectly suits humanity's needs.*

These life systems loaded Earth's crust with sufficient fossil fuels and other biodeposits to catapult humans toward a technologically advanced civilization.[32]

     The history of life on earth has one purpose, Ross believes, and that is to prepare the earth for the arrival of our current technologically-adept civilization. He states that God's purpose was to "supply physical resources for the rapid development of civilization and technology and the achievement of global human occupation."[33] Each species, Ross believes, was uniquely created by God for an express purpose related to preparing the earth for the human technological age. Hugh Ross has a very large following among evangelicals. They reject Al Mohler's young earth as anti-scientific and think that by moving to Ross' old earth view they have moved towards greater scientific respectability.

     Similarly evangelicalism has bought firmly into the ideas of Michael Behe which virtually no faculty member in any biology department of any research university in the world would endorse. Evangelicals, however, largely accept his grandiose statement as he draws the book, Darwin's Black Box to a conclusion:

The result is so unambiguous and so significant that it must be ranked as one of the greatest achievements in the history of science. The discovery rivals those of Newton and Einstein, Lavoisier and Schrodinger, Pasteur, and Darwin. The observation of the intelligent design of life is as momentous as the observation that the earth goes around the sun or that disease is caused by bacteria or that radiation is emitted in quanta.[34]

     So this is the anti-scientific world of evangelicalism and yet if evangelicals are going to truly appreciate the beauty of God's creation, they will need to pay attention to what scientific tools tell us about God's world. How can a bridge be built from science to a world like that?

     Building a bridge begins by recognizing that both sides are on firm ground. There will be no bridge to evangelical Christianity if those building the bridge believe that one side is nothing more than a figment of the imagination. There must be respect for the fact that both sides are grounded in reality and that neither side is floating free.

There must also be respect for the fact that the reason why evangelicals struggle so mightily with science has a logical basis. We evangelicals draw our purpose in life from our belief in the Bible as the Word of God. There are very strong logical reasons for doing so. We evangelicals raise our children to recognize, and want our grandchildren to be raised to realize, that the force of sin is real and not imaginary, and that salvation from sin comes through repentance followed by faith in the God who became man. We evangelicals believe it is possible to enter into a personal relationship with that God through prayer and by living in unity with the One who created all things and in whom all things are held together. There is no other way to build a bridge between science and evangelicals than for those who are building the bridge to be committed, with the deepest respect, to the values on both sides.

What Will the Bridge Look Like?

Building bridges requires plans, and organizations to implement those plans. In this case, however, the task is enormous, perhaps even more complex than that of building a physical bridge. On one side of the bridge we have that which studies the natural and has come up with conclusions that seem to completely contradict what has become dogma on the other side. The other side lives daily with the supernatural. It is used to believing the seemingly impossible. That is my world—prayer, heaven, the Holy Spirit, miracles, resurrection, and virgin birth. We evangelicals think those things are not impossible. They are, we believe, true. Why should we hesitate to believe in a young earth, when we can accept things which seem much more impossible than that? Science tells us that resurrection is impossible, does it not? Science tells us that miracles cannot take place, does it not? Doesn't science say that prayer doesn't work? The answer to all of these questions is a firm "no." Science is silent on singularities. Whether the Creator could, at one point, have changed the rules by which universe operates thereby superseding death cannot be investigated by science. Science only studies that which occurs with regularity. Science, in contrast to the common evangelical view is not wrong on issue like miracles, eternal life and the Holy Spirit—it is simply silent. Evangelicals do not understand that yet, and I don't think all scientists understand it either. Providing the Church with education about the limits of science is of the utmost importance.

What people must come to understand, however, is that there are some things about which science does enable us to be quite certain. The tools of science do tell us with reasonable certainty— details about past events. Just as a detective can put together all the clues to tell with virtual certainty who committed a particular crime, so the tools of science eventually converge with virtual certainty on what took place in the distant past. This is much different than the question of whether there is a Holy Spirit or whether there can be singularities in which the supernatural supersedes the natural. Part of bridge-building is to help evangelicals come to understand how science works. They will need to understand that just as we can be quite certain that oxygen molecules are carried by hemoglobin or that the atom consists of electrons, protons, and neutrons, so we can be equally certain that the creation of all of life has occurred through the evolutionary process. People don't understand that the tools of science bring us to near certainty regarding some matters. Science is not just having an opinion about something—science is a way of knowing that something is true. We need to help people understand this and that process needs to begin at a young age. Instituting the process will be slow and will require patience. It will need to take place in churches, Sunday Schools and Christian schools. People will need to be reassured that tenets of evangelicalism don't stand or fall by what science says. Science is God's tool and when administered correctly evangelicalism has nothing to fear since science is simply a way of knowing truth and all truth is God's.

Teaching science in ways that help Christians really understand its bounds, but also its power is essential. Developing web resources, especially film is going to be essential. Getting science-training into seminaries, workshops for pastors, curriculum for Christian schools and Sunday Schools, professional development opportunities for Christian high school teachers are all important. The issue is so all-encompassing and the misunderstanding so great, that nothing short of a highly concerted effort is needed to help evangelicals come to peace with the power and reliability of the scientific process.

However, this just raises a new question. How could it come about that seminarians will come to see the value of learning about science? Why should Christian school teachers attend a workshop that helps them understand science better? What about pastors? They are busy preparing sermons and counseling people dealing with high stress issues like broken marriages, debilitating disease, death of a parent or even a child. Why should they also take the time to attend workshops or read books that help them understand the scientific process better? So the biggest issue is not so much figuring out what needs to take place to build a bridge as it is helping people to understand why they should want to take the time from their busy lives to walk on that bridge. Helping people understand that is at least as important as building the bridge itself.

Coming to Understand the Value of Crossing the Bridge

When one is convinced that there is more to reality than the natural world we see around us, it is far too easy to dismiss that which doesn't quite fit into our supernatural world view; it is far too easy to simply say we'll understand it better by and by and refuse to think of it any further. This is even more true when one's worldview is extremely comforting. What most people are interested in learning is that which will reinforce their source of meaning, security and joy. They want to learn more about how to function effectively within that zone of comfort. It's just not clear to such individuals why getting up on a bridge and crossing over into a foreign land—a territory grounded firmly in nature and the study of nature— is a journey worth taking.

There is another factor which limits the number of people crossing the bridge. Those who know that the bridge can be built and can cross it freely sometimes are content to ignore those on the supernatural side as they live within in their zone of comfort. "Why should we want to disturb their view of nature?" I have been repeatedly asked. Since the task of getting people to walk on the bridge is so difficult and since they are quite comfortable living in their isolation, "Why should anyone want to disturb them?" I've known many people who feel that way. Since I do care about this issue, since I do want people to change, I have at times been labeled a crusader. "Just let them be," I've been told, "Don't rock the boat." That attitude propagates our current situation.

So why should we care about getting people onto the bridge?

I care because I find the Christian life immensely fulfilling and, more specifically, because I personally find that the evangelical approach to Christianity provides life with a richness which I do not want to see die, or even become diluted. Life in Christ is, as I see it, truly beautiful and, more than that, it is a journey towards truth. It is a journey that leads to where we all want to go.—a journey toward ultimate Reality. So that's why I want people to get up on the bridge and to start walking.

There are three reasons why I think evangelicals can come to see the importance of moving onto the bridge and starting the journey:
1. They will come to see that their children and grandchildren who leave their land to study in the universities which are found in that other land, will discover that science really does teach contrary facts that are highly reliable. The children, without a bridge, will stay in that other land. This experience is far too prevalent already, and as the scientific picture deepens and broadens, it won't get any better. Evangelicals need to see this for the sake of their children.
2. They will come to see that a life grounded in a scientific reality enriches, rather than detracts from, their life of faith. They will see that reading the book of nature enhances their understanding of the God they love.
3. Like the Psalmist of old, they will come to see that using science to understand God's world enriches their worship.

I would like to elaborate a little on the final point. Jesus said that if God's people keep quiet about the glory of God, the stones will cry out. The more we learn about nature, the more we hear nature crying out, "Glory." The more evangelicals learn about this nature, the more they are seeing that life on earth is not simply a passageway through a sterile secret tunnel into eternity. As they come to see this from a distance—from the other side of the bridge—they will come to desire the journey which takes them over the bridge to where the scientific tools are exploring God's glory in magnificent detail. They won't want it to be only nature that shouts out "Glory" anymore. They will want to join the chorus and they will come realize that that will entail crossing the bridge.

That is why books like this one are so important. That is why organizations like BioLogos are important. We all exist not just to construct the bridge but to show people why crossing the bridge is an all-important endeavor.

For our part in BioLogos, we will continue to supply as best as we can the theological, philosophical and scientific resources which are all part of the bridge infrastructure. We need to provide the cerebral content which will show there need be no cognitive dissidence any longer. However, while we are doing this, we will ensure that worship is pre-eminent. Nature, far from causing us to lose sight of God, leads us into God's presence. Nature cries out from the other side: "Cross the Bridge, cross the bridge." We need to facilitate the voice of nature as it calls across from the other side. We need to strengthen its vocal cords. We need to publish its poetry and we need to let it sing.

You will go out in joy
and be led forth in peace;
the mountains and hills
will burst into song before you,
and all the trees of the field
will clap their hands...
This will be for the LORD's renown,
for an everlasting sign,
which will not be destroyed." [35]

There is one primary reason why I have become convinced during this past year that worship of God through meditation on the glory of God's creation will be the key way of drawing both sides onto the bridge together. BioLogos held two workshops this past year, each of which included a mixture of perspectives about evolutionary biology. The first was a workshop for leaders of evangelicalism: leading pastors, leading scientists who hold a Christian perspective, and leading theologians and other scholars.

The title of the meeting was "In Search of a Theology of Celebration." We always knew it was an awkward title. What is a Theology of Celebration? What does it mean to search for this theology? The Organizing Committee, which included some of evangelicalism's most noted leaders: Philip Yancey, Andy Crouch, Tim Keller, Os Guinness, and at the start, Francis Collins, wanted to make it clear that no matter what our perspective on the science/faith divide, we all ought to be able to celebrate nature, in worship together when faced with the majesty of creation. So we made worship our central theme. Andy Crouch led us in worship. We sang together with one voice and we read Scripture. No matter what our perspective, we could all hear the voice of God speaking to us through nature and through each other. I have written about what I think we all sensed as the meeting came to a close and we sang: "How Firm a Foundation a foundation ye saints of the Lord, is laid for your faith in his excellent Word!" We had all met on the bridge, each with our different perspectives, and it had not collapsed.[36]

Several months later we held a workshop for biology teachers who work in Christian schools. We discussed the data which supports evolution. About half of the teachers present had a young earth perspective and a significant number of the remainder had been heavily influenced by the anti-Darwin component of the Intelligent Design Movement. This experience was also deeply embedded in worship of the God of Creation. We all had that in common. That workshop with a group of biology teachers, many of whom thought so differently than me about evolution, was among the most enriching experience of my life—and again it was worship which brought us together. With tears streaming down some of our faces, we stood on the not-yet-completed bridge together, and we, a group of biologists, sang of God's faithfulness as we meditated on the glory of God revealed through God's creation.

Truly we can come to peace with science only as we listen to the voice of God calling us onto the bridge, where we will meet together to hear his voice spoken through the words of nature:

The heavens declare the glory of God;

the skies proclaim the work of his hands.
Day after day they pour forth speech;
night after night they display knowledge.
There is no speech or language
where their voice is not heard.
Their voice goes out into all the earth,
their words to the ends of the world.[37]

With that, nature will not need to cry out in the absence of the evangelical voice any longer. We together will join nature and, standing on the bridge, will call out in one voice, "Glory."

From the Editor...
   Let us stop for a moment and review our journey thus far. We began with a sense of awe that stirs within us as we gaze into the vastness of the universe, led on that journey of discovery by astronomer Jennifer Wiseman. We then observed what happens to that sense of awe when we begin to put language around it. This led to a discussion of the role of rhetoric, as offered in the essay by David Thomas. We continued to consider the role of language as we entertained whether we should strive to find unity in language, or could we unite on goals and purposes rather than particular narratives and language. This was explored in the essay by Cybelle Shattuck. We then recognized that although different religious narratives may be able to avoid integration issues, the one relationship that does seem to cry out for integration is that between faith and science. We noted that this is a particular challenge for evangelical Christians, as a self-

proclaimed evangelical, Darrel Falk, effectively illustrated in his own personal story.

We are on journey to explore the sanctity of nature, and the human place within it. In order to go further, we must adopt a language to express our sense of the origin of nature, and, of humanity's origins within nature. Here, then, it is necessary to state that the narrative that we will adopt is that of theistic evolution. It is our conviction that evolutionary science is enlightening our sense of how and when nature came to be in its present form. For some, the introduction of evolutionary processes diminishes the sense of a sanctity to nature, because evolution seems to distance God from an active role in nature's development. In addition, evolutionary theory seems to resist ideas of progress or of design, thus stripping nature of any inherent purpose. In light of these issues, is it possible to adhere to the theory of evolution and believe in the sanctity of nature? This is the question at the heart of our next essay, written by Dr. Denis Alexander. Dr. Alexander is the Director of the Faraday Institute for Science and Religion at Cambridge, where he is a Fellow. Dr. Alexander was previously Chairman of the Molecular Immunology Programme and Head of the Laboratory of Lymphocyte Signalling and Development at the Babraham Institute, Cambridge. In this essay, he asks an important question: Is Evolutionary History Sanctified? (This essay is a modified version of a talk given at the Darwin Festival, University of Cambridge, July 6th, 2009.)

Chapter 5
Is Evolutionary History Sanctified?

by Denis R. Alexander

Evolution presents an ambiguous picture to those exploring the sanctity of nature. On one hand evolutionary history has delivered the impressive and beautiful array of biological diversity that presently exists on planet earth, not to speak of those millions of extinct species that we can reconstruct in our imaginations from the fossil data. Evolution has brought diversity, complexity, cooperation, creativity, intelligence, communication and, more recently, language, conscious awareness and moral responsibility. On the other hand, evolutionary history involves predation, pain, suffering and death on a vast scale; the positive outcomes have come, and continue to come, at great cost. The genetic variation without which life would not exist is also the same variation that guarantees disease. Evolution speaks with an ambiguous voice.

In light of such ambiguity, can we refer to evolutionary history as being 'sanctified'? 'Sanctity' refers to 'the state or quality of being sacred or holy', but if the notion is to have any traction in the world of evolutionary biology, then some initial groundwork needs to be carried out to address related questions to do with Progress and Purpose. The biological world could still be 'sacred or holy' and yet not, in its evolutionary history, display any particular signs of inevitable progress. But if it had no purpose at all in any ultimate sense, then at the very least the notion of it being 'sacred or holy' might be difficult to sustain. The sanctity of nature might in that case arise from a temporary assignment of value to it by humankind during that very brief moment that humans have existed on planet earth, but without some broader narrative the notion of 'sanctity' certainly loses much of its content. To explore these ideas further requires an investigation into historical and contemporary discussions on the question of progress and purpose in evolutionary history.

Progress

The idea of evolution was born in a cultural context in which the idea of progress was a dominant theme, but the discussion about whether evolution does in fact display at least some kind of progress continues right up to the present day. In fact the topic interested the late Stephen Jay Gould so much that his swan-song, his 1,433 page book The Structure of Evolutionary Theory, published in 1992, the year of his death, largely revolves around this vexed issue.

Evolution as natural history began with Jean-Baptiste Lamarck (1744-1829) who believed in the fixity of species until 1797 but who then became an evolutionist, as he stated in his introductory lecture for his new post as Professor of Lower Animals at the newly founded Natural History Museum in Paris in 1800. Lamarck followed this up by three major publications in which he presented evolution in strongly progressionist terms, but ironically Lamarck himself, an ardent materialist, was also a convinced uniformitarian. To bring these two apparently incompatible ideas together, Lamarck envisaged the continuous spontaneous generation of new species which then move up the escalator of life, with all steps occupied at all moments. This is the primary process and it is then the differing circumstances on each step that lead to different adaptations and consequent variations. But for Lamarck it is the great escalator of evolution which is the main story moving all of life in an upward direction.

Darwin's own Grandfather, Erasmus Darwin (1731-1802), poet, rationalist, botanist, to some degree foreshadowed Lamarck, in his work Zoonomia (1794-6), envisaging that all living animals had arisen millions of years before man from one "living filament" which the great First Cause had endowed with the potential for delivering "improvements by generation to its posterity, world without end!" But it was Lamarck who was the real natural historian who put the idea of evolution on the scientific map.

Lamarckian themes were picked up by the Scottish publisher Robert Chambers in his then anonymous Vestiges of the Natural History of Creation (1844). Chambers presented his readers with a developmental hierarchy, which he termed the "universal gestation of nature". It was basically a story of the evolution of everything. In the sky a swirling fire-mist evolved into nebulae, solar systems, and planets; on the ground invertebrates, fish, reptiles, mammals and man followed in order up life's great escalator; and in society there was development in civilization as Negro, Malay, American, Indian, Mongolian and Caucasian gave way one to the other. The book was a sensation and it was not until the 1890s that the sales of Darwin's Origin of Species began to catch up with Chambers' popular work, despite, or perhaps because of, the Vestiges being lambasted by all the leading natural philosophers of his time.

If we were to score Lamarck, Erasmus Darwin and Chambers 9 or 10 on a scale of 1-10 in the Progressionist stakes, then I suspect that Charles Darwin himself would score around 5. As always he was temperate in his comments, at least by the standards of his time, balancing one comment off with another. On one hand in the Origin of Species we find Darwin writing that: "as natural selection works solely by and for the good of each being, all corporeal and mental endowments will tend to progress towards perfection". Progress for Darwin was a consequence of biotic competition. In crowded ecosystems full of competing life forms, the constant removal of inferior by superior life forms would impart a progressive direction to evolutionary change in the long run. But then Darwin writes in his letter to the American progressionist palaeontologist Alpheus Hyatt on Dec 4th 1872: "After long reflection I cannot avoid the conviction that no innate tendency to progressive development exists". And we find Darwin scribbling in the margins of a progressionist text: "Never say higher or lower".

Darwin's most enthusiastic supporters simply dispensed with his caution and propounded a robustly progressionist view of evolutionary history. Herbert Spencer, arguably the most famous philosopher of his age, had already started working out a great developmental Lamarckian scheme for the evolution of nearly everything in 'Progress: Its Law and Cause' published in 1857 and simply absorbed bits of Darwinism into his scheme as they came along, but remained more Lamarckian than Darwinian for the rest of his life. Spencer maintained that the end point of the evolutionary process would be the creation of 'the perfect man in the perfect society' with human beings becoming completely adapted to social life. Darwin didn't use the word 'evolution' at all in the first edition of the Origin because it carried the sense of 'unfolding' with a strong connotation of inevitable progress, but Spencer pushed the word heavily and Darwin first started using it in The Descent of Man in 1871.

The progressionist tradition continued on unabated right through the 20th century, with evolutionary thinkers in the first half, such as T.H.'s grandson Julian Huxley and R.A.Fischer, together with the Catholic Lamarckian palaeontologist and theologian Teilhard de Chardin, in their very different ways keeping alive a progressionist stance. But as Michael Ruse points out, following the development of the neo-Darwinian synthesis in the 1920s and 30s, it now became much less respectable to talk about progression in scientific publications. Instead such material was relegated to the popular writings of the evolutionary biologists, as in Julian Huxley's hugely prolific output during the 1920s to 1950s, Huxley being attracted to vitalism and the writings of Henri Bergson.

Many of the great evolutionary biologists of the latter half of the 20th century, such as Ernst Mayr and E.O.Wilson were likewise convinced progressionists, Wilson writing that "Progress, then, is a property of the evolution of life as a whole by almost any intuitive standard, including the acquisition of goals and intentions in the behaviour of animals".

Not until the writings of Stephen Jay Gould do we find a really vigorous all-out onslaught on progressionism, Gould proclaiming in 1989 that it is a "noxious, culturally embedded, untestable, non-operational, intractable idea that must be replaced if we wish to understand the patterns of history". According to Gould, we are a "momentary cosmic accident," albeit a "glorious accident." Summing up his view, Gould writes: "Wind back the tape of life to the early days of the Burgess Shale; let it play again from an identical starting point, and the chance becomes vanishingly small that anything like human intelligence would grace the replay".

And yet even Gould seems to have moderated his position in later life in The Structure of Evolutionary Theory and 1433 pages really does allow an author to add plenty of "ifs", "ands" and "buts". For example, on p. 468 we find Gould commenting: "But the history of life includes some manifestly directional properties – and we have never been satisfied with evolutionary theories that do not take this feature of life into account". It turns out that the progressionist accounts that Gould liked to attack most heartily were, if not exactly windmills, then at least items that can leave other forms of progressionist narrative untouched.

So we find Richard Dawkins coming across as rather a strong progressionist in his 1997 review of Gould's book Full House - "progress to mean an increase, not in complexity, intelligence or some other anthropocentric value, but in the accumulating number of features contributing towards whatever adaptation the lineage in question exemplifies. By this definition, adaptive evolution is not just incidentally progressive, it is deeply, dyed-in-the wool, indispensably progressive". And in his Ancestors Tale we find Dawkins writing that "the cumulative build-up of complex adaptations like eyes, strongly suggests a version of progress — especially when coupled in imagination with some of the wonderful products of convergent evolution." Dawkins also sees the evolution of evolvability itself as a second type of progressive narrative, writing that "there really is a good possibility that major innovations in embryological technique open up new vistas of evolutionary possibility and that these constitute genuinely progressive improvements".

This brief thumb-nail sketch reminds us of the immensely long and complex debate that has surrounded the whole issue of progression in the evolutionary literature. And the question I now wish to ask is whether the outcome of this discussion – that is, whether evolutionary history displays progression or not - makes much difference to the question of the sanctity of nature, particularly within the framework of Christian theology, and the answer I would like to give is "not much".

The traditional Christian understanding of evolution, which started with thinkers like Charles Kingsley and Frederick Temple as soon as the Origin was published, and continues in an unbroken lineage since that time, is that it represents the creative process that God uses to bring about his intentions and purposes for biological diversity in general and for humankind in particular. Therefore every twig of every branch on the great evolutionary tree of life is interconnected and plays its role in the historical narrative. Every twig has an intrinsic value that is bestowed upon it by God, as much the 99% of species that have gone extinct as the ones that remain alive. God cares for and enjoys every aspect of his creation in its own right and by its own criteria, not just for the utilitarian reason that each species might be the ancestor of something different (1 Chron. 16: 23-34; Job 38; Psalm 104). The created order resounds with praise to God (Psalm 148) not only because it reflects God's power and wisdom in creation, but also because each aspect both great and small (Matt. 10:29) is valued in itself.

This is why, from the perspective of Christian theology, the outcome of the discussion about evolutionary 'progression' in the literature, makes little if any difference to the question of the sanctity of nature. One does not have to believe in some internal progressivist principle within the evolutionary process in order to be able to acknowledge the intrinsic value of its every aspect, both now as in the past.

Of course, this is not to say that a valid progressionist story cannot be told. Stand back and look at evolutionary history taken as a whole and the arrow of evolutionary time is inescapable. As Sean Carroll from the University of Wisconson-Madison remarks in a recent review in Nature: "Life's contingent history could be viewed as an argument against any direction or pattern in the course of evolution or the shape of life. But it is obvious that larger and more complex life forms have evolved from simple unicellular ancestors and that various innovations were necessary for the evolution of new means of living". If we take the criterion of complexity as a read-out for progression, then there is no doubt that evolutionary history encompasses progression. And as we have seen, for evolutionary biologists such as Dawkins, adaptation during evolution "is not just incidentally progressive, it is deeply, dyed-in-the wool, indispensably progressive".

My point here is not that these insights are not valid – personally I think they are – but rather that even if they were not, the sanctity of nature as the realm of God's created order would still remain the case. That conclusion is independent of the outcome of arguments about evolutionary progression.

Purpose

What about the other 'P' word hovering behind Progression: Purpose? It is quite possible to be an ardent progressionist in evolution, without believing that the process taken as a whole has any ultimate purpose. This is clearly the position of Dawkins when he writes, perhaps on a rainy day in Oxford: "The universe we observe had precisely the properties we should expect if there is, at bottom, no design, no purpose, no evil and no good, nothing but blind pitiless indifference."

The philosopher Daniel Dennett agrees – Dennett asks whether the complexity of biological diversity can "really be the outcome of nothing but a cascade of algorithmic processes feeding on chance? And if so, who designed that cascade?" Dennett answers his own rhetorical question by saying: "Nobody. It is itself the product of a blind, algorithmic process". "Evolution is not a process that was designed to produce us".

The discussion here is not then about the question of progression as such, but about the question of purpose, the word that Dawkins uses. And here we do have to mark a parting of the ways. For clearly the idea of purpose is implicit in the kind of theological outline that I have introduced. The idea is that indeed the evolutionary process is fulfilling God's creative intentions and purposes. But on the other hand I rather agree with Dawkins and Dennett that if you look at the evolutionary process as an atheist and simply through the window of biology, then there is nothing there that forces upon you a narrative of ultimate purpose.

So what does everyone do in practice in light of the metaphysically ambiguous record provided by evolutionary history? Well of course everyone imposes their own narrative of purpose upon their lives in the context of the world around them, for the sake of mental health, if no other. In practice humans really can't live without some kind of meaning and purpose.

For writers such as Dawkins it is the metanarrative provided by an ultra-Darwinian world-view which provides the framework within which purpose is found. The understanding of and participation in the evolution of the universe in general, and life on planet earth in particular, provides its own reward. Those of a more existential frame of mind turn their faces bravely towards the meaninglessness of the universe and courageously proclaim their human decision to will their own purpose in the teeth of its darkness, perhaps in the pursuit of a successful career.

My purpose here is not to critique any of these positions, and of course there are many more, but simply to point out that the metaphysical imposition of purpose upon the evolutionary narrative by the theologian is not so different from the metaphysical imposition of other kinds of purpose that everyone employs anyway. None of them without any exception can actually be derived rationally and bottom-up from the evolutionary process itself.

So does that leave advances in our understanding of evolution completely divorced from our theological understanding of purpose, with the implications that has for our understanding of the sanctity of nature? Not necessarily. We cannot in my view derive theology from the evolutionary process itself, though some have tried to do just that, but I do think that our current understanding of biology renders less plausible the suggestion that evolutionary history necessarily lacks any plan or purpose. Scientific theories are viewed more favorably when they make predictions that can, in principle, be falsified. It is therefore interesting to note that it is the nature of the evolutionary process itself that can be cited to count against the claim, made by writers such as Dawkins and Dennett, that it is necessarily without purpose. Let us consider five examples that illustrate this point, taking care to emphasize that none of these examples in themselves indicate that evolutionary history has some purpose but, taken together, present counter-evidence to the claim that evolutionary history is of necessity devoid of purpose.

First and most obviously evolution taken as a whole is not a chance process. This at least is where atheists and theologians can sing from the same song sheet. As Dawkins writes in 'The Blind Watchmaker': "One of my tasks will be to destroy this eagerly believed myth that Darwinism is a theory of 'chance'". Of course the process incorporates chance in the generation of genomic variation, of course there are stochastic events leading to mass extinctions, but the winnowing effect of natural selection ensures that in the finely tuned interaction between chance and necessity it is necessity which wins in the end. So evolution is a highly organized and in many ways highly conservative process.

Second, stand back and look at the 3.8 billion years of evolution as a whole and, as we have already observed, the striking increase in biological complexity is obvious. For the first 2.5 billion years of life on earth, things only rarely got bigger than 1 millimetre across, about the size of a pin-head. There were no birds, no flowers, no animals wandering around, no fish in the sea, but at the genetic level there was lots going on, with the generation of many of the genes and biochemical systems that were later used to such effect to build the bigger, more complex and diverse living things that we see all around us today. At the same time the oxygen levels in the atmosphere increased to the point at which more complex life-forms could be sustained.

It was not until the advent of multi-cellular life from around perhaps one billion years ago that living organisms start to get bigger, although even then they were generally on a scale of millimetres rather than centimetres. Only in the so-called 'Cambrian explosion' during the period 505-525 million years ago do we find sponges and algae grow up to 5-10 cm across, and the size of animals began to increase dramatically from that time onwards, until today we have creatures like ourselves with our brains with 1011 neurons with their 1014 synaptic connections or more, the most complex known entities in the universe.

Third, underlying biological complexity are networking principles that are turning out to be fewer and simpler than they might have been, pointing to constraint and elegance, not to randomness and the idea that 'anything goes'. Given that in every cell complex networks of interactions occur between thousands of metabolites, proteins and DNA, this is quite surprising. As Uri Alon from the Weizmann Institute comments: "…Biological networks seem to be built, to a good approximation, from only a few types of patterns called network motifs"…."The same small set of network motifs, discovered in bacteria, have been found in gene-regulation networks across diverse organisms, including plants and animals. Evolution seems to have 'rediscovered' the same motifs again and again in different systems…"

Fourth, the very limited array of protein structures used by living organisms compared to the astronomically huge number of possible structures is also very striking. If you look at all the known proteins in the world, and their structural motifs, based on all the genomes that have been sequenced so far, you find that the great majority can be assigned to only 1400 protein domain 'families'. In other words, all living things are united not only by having the same genetic code, but also by possessing an elegant and highly restricted set of protein structures.

The concept of 'fitness landscapes' can be applied to protein structure and function. Again and again it turns out that the evolutionary pathways to arrive at a particular function of a particular protein (such as an enzyme) are remarkably constrained. In other words, there are only a few ways to arrive at a particular protein function because only some genetic mutations will get you there and not others. It is as if an evolutionary path is laid out in front of the gene encoding the enzyme, and the genetic dice keeps being thrown until the enzyme structure is generated that optimises fitness for its particular function. This is no random process, each step along the way being preserved by benefits to the organism that uses the enzyme. As the authors reviewing such recent findings conclude: 'That only a few paths are favored also implies that evolution might be more reproducible than is commonly perceived, or even be predictable', and as another recent paper on this topic concludes: "We conclude that much protein evolution will be similarly constrained. This implies that the protein tape of life may be largely reproducible and even predictable".

Overall it appears that around 98% of all the amino acids in all proteins cannot change because of the striking decrease in fitness of the organism that would result. This means that the genes that encode these amino acids cannot change either, at least not by mutations that change the amino acid sequence. This might sound like a recipe for a static protein world. In practice this is not the case: proteins do evolve, but they just do so really slowly and cautiously. For example, if other random mutations occur in the same protein, then the constraint on the 98% of 'frozen' amino acids is lifted somewhat. It's unlikely that the evolutionary search engine has yet completed its job of searching the complete repertoire of protein 'design space', but it has come a long way in 3.8 billion years, and the present 'snap-shot' that we have certainly points to a highly constrained molecular world. In practice what this means is that if you generate a random jumble of amino acid sequences, the vast majority (indeed an astronomically huge number) will have no function at all, and it is up to the evolutionary search engine to find the tiny number that have functions useful for life.

Fifth, there is the remarkable phenomenon of convergence, the repeated evolution in independent biological lineages of the same biochemical pathway, or organ or structure, to which writers such as Dawkins and Simon Conway Morris, have drawn repeated attention. At the phenotypic level these can be very striking. The hedgehog tenrecs of Madagascar were long thought to be close relatives of 'true' hedgehogs, because their respective morphologies are so similar, but it is now realized that they belong to two quite separate evolutionary lineages and have 'converged' independently upon the same adaptive solutions, complete with spikes. The convergence of mimicry of insects and spiders to an ant morphology has evolved at least 70 times independently. The technique of retaining the egg in the mother prior to a live birth is thought to have evolved separately about 100 times amongst lizards and snakes alone. Compound and camera eyes taken together have evolved more than 20 different times during the course of evolution. If you live in a planet of light and darkness, then you need eyes – so that's what you're going to get!

Evolution is a search engine for exploring design space. Biological diversity is definitely not a case of "anything can happen". Only some things can happen, not in a deterministic way, but in a highly constrained way. So far from looking stochastic and random, evolution looks highly organized and constrained.

None of this entails that some overall purpose but can be 'read off' the evolutionary narrative. But it does, I think, act as evidence against the claim that evolutionary history is necessarily without purpose. For it is a characteristic of purposeless narratives that they meander, they lack constraint, stochastic elements are dominant and in the end anything goes. But in reality evolutionary history is highly organized and constrained – even directional - which certainly seems to be at least consistent with the theological claim that there is a God who has intentions and purposes for the world.

Purpose and Suffering

If there is an overall purpose in evolutionary history, and it is not simply a "tale told by an idiot", then this casts a very different light on the problem of the pain and suffering that is intrinsic to the evolutionary process. There is a necessary cost to the process; without that cost no living organism would exist. Everything exists in great food-chains. We are all dependent on each other. The dead constantly make way for the living. And biological life itself is a 'package deal' in which every single beneficial aspect – beneficial that is from the perspective of the well-being and evolutionary fitness of the individual organism – also has a downside, a negative aspect that the organism would prefer not to have.

We can build a great Table, with two columns, in which all the 'pluses' lined up on one side are finely balanced by all the 'minuses' on the other. Pain is essential to our survival and so is a necessary good. Eating food seems a pretty good plus, although it has a minus: it increases the dangerous form of oxygen ('free oxygen radicals') inside cells, so increasing the chance of DNA damage. All cells (that have a nucleus) are programmed to self-destruct should that be necessary, by a process known as 'apoptosis'. This self-destruction is essential and positive during the development of the brain, nerves, muscles, and many other organs. Conversely, when the process of apoptosis becomes dysfunctional, then it can lead to cancer. The list in the Table is endless.

Such insights are in themselves insufficient to provide fully satisfactory answers to the challenge of suffering. But they are a reminder that once some overall purpose in the evolutionary narrative is established, then at least the costly nature of the narrative makes more sense. We are all familiar with the idea that achieving certain goals can be costly, and we also believe that the cost can be worth it if the goal is worth it. Collectively the whole evolutionary tree of life only exists because each tiny leaf and twig bears a small part of the collective cost necessary for its existence. It is a sobering thought.

Purpose and the Sanctity of Nature

If the process of evolution is not incompatible with some larger narrative of purpose, which it clearly is not, then likewise it can be positively compatible with the kind of narrative provided by Christian theology. As already mentioned, everyone without exception imposes their own metanarrative upon evolutionary history, be it the "no design, no purpose, no evil and no good" metanarrative of Dawkins, the "universal acid" of Daniel Dennett, or the metanarrative of creation care and future hope provided by Christian theology.

For the sanctity of nature in Christian theology is embedded in eschatological hope. The created order has intrinsic value now. It has the 'quality of being sacred or holy' because it has been 'set apart' by God for his intention and purposes, to be cared for by humankind 'made in his image'. That care is to be exerted now for our own present good and that of the generations to come. This point was brought out powerfully by Calvin when he wrote in 1554 AD in his Commentary on Genesis that: 'The earth was given to man with this condition, that he should occupy himself in its cultivation.......The custody of the garden was given in charge to Adam, to show that we possess the things that God has committed to our hands, on the condition, that being content with frugal and moderate use of them, we should take care of what shall remain...Let him who possesses a field, so partake of its yearly fruits, that he may not suffer the ground to be injured by his negligence, but let him endeavour to hand it down to posterity as he received it, or even better cultivated... Let everyone regard himself as the steward of God in all things which he possesses. Then he will neither conduct himself dissolutely, nor corrupt by abuse those things which God requires to be preserved".

Sacred nature deserves to be cared for by good and careful stewards, but there is more than just caring for it in order to hand down in better shape to posterity. Within the Christian framework there is also the expectation of a new heavens and a new earth yet to come (Isaiah 65:17; 2 Peter 3:13; Rev 21:1), providing the key ultimate goal and purpose for the present created order. The new heavens and the new earth involve both continuity and discontinuity, a transformation that incorporates all the best aspects of the present state into that which is to come. Whereas the details in the tableau are sketchy, the implications for us are clear: the sanctity of nature has an eternal aspect that impacts in a very practical way on our own responsibilities for caring for planet earth in the here and now. What we do to the planet has eternal consequences. This is where the scientific story for the future of the universe and the theological story part company. The current scientific consensus for the end of the universe is ultimate boredom, a heat death in which there is no longer sufficient energy for anything interesting to exist. If that is the only story, then the universe has no ultimate purpose and therefore any notions of the sanctity of nature must be constructed on very transient foundations. But once eschatological hope is incorporated into the story, then the idea of the sanctity of nature becomes endowed with a powerful underpinning that provides a potent motivation for present creation care. Eternal things are worth looking after.

The ultimate purpose incorporated within the new heavens and the new earth also helps to render coherent the existence of the costly evolutionary tree of life. Evolutionary history is sanctified, set apart, to fulfill God's good purposes for the universe, for really good things come at a cost.

BIBLIOGRAPHY

Alexander, D. Creation or Evolution – Do We Have to Choose? Monarch, 2008.

Alexander, D. & White, R. S. Science, Faith and Ethics: Grid or Gridlock?, Hendrickson, 2006.

Morris. S.C. Life's Solution – Inevitable Humans in a Lonely Universe, CUP, 2004.

Morris, S.C. The Deep Structure of Biology: Is Convergence Sufficiently Ubiquitous to Give a Directional Signal? Templeton Foundation Press, 2008.

Spencer, N. and White, R.S. Christianity, Climate Change and Sustainable Living, SPCK, 2007.

White, R.S. (ed) Creation in Crisis, SPCK, 2009.

From the Editor...

In the previous essay, Denis Alexander began with the question of whether evolutionary history can be sanctified. He ended his essay with an affirmative answer. Adopting an evolution narrative for the origin of nature can, indeed, lead us to a belief in the sanctity of nature, and of our human place and role within it. But can we go further? It is one thing to ask whether a belief in evolution can accommodate a sense of nature's sanctity, but quite another to suggest that the evolution narrative is the one that truly leads us to a higher sense of the sanctity of nature, and a more appropriate sense of our human condition within it. Dr. Michael Zimmerman makes this case in our next essay.

Dr. Michael Zimmerman is the founder and director of The Clergy Letter Project, an international organization of religious leaders and scientists created to demonstrate that religion and science need not be in conflict. He has been teaching biology at Butler University in Indianapolis, but has recently accepted the position of Vice President for Academic Affairs at The Evergreen State College in Olympia, Washington State.

Chapter 6
Why an Understanding of the Evolution/ Creation Controversy is Essential for a Robust Appreciation of Nature

by Michael Zimmerman

In 1973 the great population geneticist Theodosius Dobzhansky published an essay in The American Biology Teacher with the provocative title "Nothing in Biology Makes Sense Except in the Light of Evolution."[38] As one of the leading forces behind what has become known as the modern evolutionary synthesis, Dobzhansky was certainly in an appropriate position to explain the importance and centrality of evolution to all of biology.

This is exactly what he did in that paper, a paper that has become a classic in the field. Two sentences from Dobzhansky's paper masterfully and succinctly define his position: "Seen in the light of evolution, biology is, perhaps, intellectually the most satisfying and inspiring science. Without that light it becomes a pile of sundry facts – some of them interesting or curious but making no meaningful picture as a whole."

Interestingly, Dobzhansky's use of the evocative phrase "light of evolution" came not from a scientist but from a religious leader, the Jesuit priest, paleontologist, biologist and philosopher Pierre Teilhard de Chardin (1881-1955).

Now, more than 35 years later, with biological knowledge having increased exponentially and molecular genetics and genomics having added incredible layers of complexity to our understanding of the natural world, Dobzhansky's quotation is more true than ever before. At the same time, his quotation is cited so frequently that I fear people no longer think about the meaning of his words. It is important, however, to understand his meaning if we are to fully appreciate our place in the natural world.

In his latest book, Geerat Vermeij, past editor of both Evolution and Paleobiology the leading journals in their respective fields, like

Dobzhansky before him, discusses the significance of evolution. In moving prose, Vermeij defines evolution as an idea so powerful, so beautiful, and so far-reaching in its implications for human existence, that every educated person can be enriched and enlightened by it. Evolution – descent with modification – is a concept that organizes, explains, and predicts a multitude of unconnected facts and phenomena of life in nature past and present. It provides a coherent framework for understanding where we came from, where we are going, and how we and the rest of living nature create bewildering complexity, a world of meaning, and surpassing beauty. Quite simply, evolution has outgrown its original home in biology and geology. It is the foundation of a worldview in which environments, genes, organic architecture, physiology, chance, the economic struggle for life, and historical narrative come together to illuminate how we live in the world.[39]

      Understanding the centrality of evolution means a number of things. First, it means that we can grasp the relationship between humans and the rest of the natural world. Indeed, we can appreciate the fact that we are, in more than just a metaphysical sense, one with the natural world rather than being separated from it. We share a common heritage with all living things and that heritage can be read in our DNA. We are now able to look carefully at genomic structure and determine just how closely related we are to other organisms – and we are able to determine when we last shared a common ancestor with any of those organisms.

      Understanding the centrality of evolution also means that we can begin to comprehend the intricate interrelationships that exist within natural communities. We can observe with new insight the complex interactions inherent in ecosystems and recognize that all species, including but not limited to humans, have played a role in shaping the environment as we've come to know it. Vermeij also addresses this point with his standard eloquence:

      Organisms do not live in a vacuum. They are independent actors, pursuing their self-interest as competitors, but they are also linked to one another in a complex web of relationships. They create a civilization of sorts, a system of antagonisms and productive partnerships in which participants are well adapted to each other, a whole that nourishes life and that thrives on adaptation. If that

*civilization is disturbed, its constituents will suffer. Humans are part
of that civilization of life. We benefit from it and we cannot escape its
embrace. Given our power and ambition to alter the world, we have
a special global responsibility to do what we can to promote that
civilization.*[40]

The third and, for our purposes here, final way that
understanding the centrality of evolution can affect our worldview is
by providing insight into the concept of emergence. In simplified
terms, emergence can be posited in contrast to reductionism. Where
the latter claims that complex systems are nothing more than the sum
of their parts and that they can be fully understood by a detailed
examination of their constituent parts, the former takes a more holistic
view. The concept of emergence argues that complex systems and
patterns actually arise from numerous , relatively simple interactions
and in so doing create something entirely novel. As Jeffrey
Goldstein, in part, described emergence in 1999 in the inaugural issue
of the journal Emergence, its common characteristics include "radical
novelty (features not previously observed in the system),"
"coherence" or "integrated wholes that maintain some sense of
identity over time," and the fact that it is the product of a dynamical
process.[41]

Stuart Kauffman is perhaps the best known of those who have
taken the concept of emergence and used it to advance our
understanding of natural systems. In his challenging book
Reinventing the Sacred, he compares reductionism to emergence:
*In this book I will demonstrate the inadequacy of
reductionism. Even major physicists now doubt its full legitimacy. I
shall show that biology and its evolution cannot be reduced to physics
alone but stand in their own right. Life, with its agency, came
naturally to exist in the universe. With agency came values, meaning,
and doing, all of which are as real in the universe as particles in
motion. "Real" here has a particular meaning: while life, agency,
value, and doing presumably have physical explanations in any
specific organism, the evolutionary emergence of these cannot be
derived from or reduced to physics alone. Thus, life, agency value,
and doing are real in the universe. This stance is called emergence.*[42]

Deny evolution and all of the above becomes meaningless. And this is exactly what creationists, in all of their guises are doing. Young Earth creationists, like those at Answers in Genesis, believe that the Earth is approximately 6,000 years old. When this short time frame is overlaid on natural communities, the possibility for the intricate interspecies relationships that are so common in nature vanish. A wider group of creationists, including young Earth creationists, deny the concept of descent with modification arguing instead that God created members of various taxonomic divisions exactly as we see them today. In addition to the horrendous problem of not being able to specify which taxa have remained unchanged, in other words, which taxa are the originally created "kinds," these creationists simply do not accept that all life forms share a common ancestry and all that entails. Finally, those creationists who promote the non-scientific concept of intelligent design, among other things, refuse to acknowledge the concept of emergence. Instead, in the face of irrefutable evidence to the contrary from all fields of science, from physics and computer science to biology and chemistry, they simply assert that complex systems cannot self-organize and must, instead, be the result of a conscious designer.

Creationists, in all guises, come to their conclusions because of the way they interpret the Bible, not because of any scientific evidence. Most recently, that intelligent design is simply an outgrowth and reformulation of earlier creationist doctrine has been ably documented by Barbara Forrest and Paul Gross in their book Creationism's Trojan Horse.[43] What is most ironic, though, is that those who demand a literal interpretation of Genesis, steadfastly argue that they are not offering any interpretation at all. Ken Ham, president of Answers in Genesis, speaks for many when he says the following:

People say you have a particular interpretation of Genesis. I don't think so. I think I just read it, and what it says is what it means. Other people interpret it and they get into trouble. That's the problem, I think. Now I believe that God created in six literal days and I believe that it's important. In fact, I believe it relates to the authority of scripture and the gospel.

Ham's comments make it clear that, regardless of what creationists might say in various contexts, the evolution/creation controversy isn't about science but rather it is about "the authority of scripture." Even ignoring the very real questions associated with that nature of interpretation, such a position is problematic on a number of levels. Most importantly, it demands that adherents forsake the hard won knowledge gained by generations of scientists when that knowledge appears to conflict with a specific scriptural interpretation. The converse of this is that adherents must willfully embrace ignorance of some of the greatest discoveries made by humans. And, by promoting a worldview that is so completely at odds with that of the world's scientific community, creationists are advocating that their static understanding of the natural world and their scientific ignorance be transmitted to future generations.

Those of us who care deeply about science literacy and who recognize that such literacy can play an important role in democratic societies, are very troubled by this position. Similarly, those of us who have a broader and far more nuanced view of religion and spirituality, are also very troubled by this position. And, I hasten to add, there are many who fall into both of the categories just mentioned.

One such group is an organization I founded called The Clergy Letter Project. Comprised of more than 14,000 clergy members and approximately 1,000 scientists from all corners of the world, this group attempts to raise the level of discourse on the topic of the relationship between religion and science. As part of the organization's activities, more than 12,600 Christian clergy in the United States have signed an open letter affirming the centrality of evolution and making it clear that nothing in evolutionary theory threatens their deeply held religious convictions. (The Clergy Letter Project also has similar letters for American rabbis and Unitarian Universalist clergy members in the States.) The powerful two-paragraph letter was written by a United Church of Christ minister and stands in stark contrast to the position staked out by most creationists. Consider the following three excerpts from The Christian Clergy Letter[44]:

Religious truth is of a different order from scientific truth. Its purpose is not to convey scientific information but to transform hearts.

We believe that the theory of evolution is a foundational scientific truth, one that has stood up to rigorous scrutiny and upon which much of human knowledge and achievement rests. To reject this truth or to treat it as "one theory among others" is to deliberately embrace scientific ignorance and transmit such ignorance to our children. We believe that among God's good gifts are human minds capable of critical thought and that the failure to fully employ this gift is a rejection of the will of our Creator. To argue that God's loving plan of salvation for humanity precludes the full employment of the God-given faculty of reason is to attempt to limit God, an act of hubris.

We ask that science remain science and that religion remain religion, two very different, but complementary, forms of truth.

The members of The Clergy Letter Project, like many other spiritual and religious individuals, thus recognize that there need not be any conflict between religion and evolutionary theory and they assert that those who promote the view that people must choose between the two are offering a false, and ultimately meaningless, dichotomy. Those who claim that science in general and evolution in particular must, necessarily, lead to an atheistic worldview are thus demonstrably wrong. In fact, although scientific knowledge certainly does not have to lead to deeper religious understanding, for some this is exactly what happens. Historically, rather than railing against advances in astronomy, geology, biology or any other science, some individuals find that the wonders of science enhance and deepen their awe and gratitude towards God.

The Clergy Letter Project also sponsors an annual Evolution Weekend celebration in which hundreds of local congregations from all over the world undertake some activity designed to educate parishioners about the compatibility of religion and science while improving the quality of the discussion about this topic. In an explicit attempt to link science and religion with a concern for the natural world, the 2011 celebration adopted an environmental theme. As the Evolution Weekend 2011 web page explains it: *The information and understanding gained through legitimate scientific inquiry can be of significant help to people of faith in better understanding this wonderful planet that we live on – its beauties and wonders, as well as the many environmental threats to the health of both natural and human communities. Science can thus be of assistance to religious leaders and communities, as they seek to fulfill their calling to care for the Earth, through more informed advocacy and actions[45]*

Good environmental stewardship, stewardship of the sort many religions advocate, requires a solid grounding in evolutionary biology. In the absence of such a sound foundation even the best intentions are likely to yield less than satisfactory results. And with strong enough creationist leanings, the intentions are not always close to the best. Journalist Bill Moyers made this case very well when he apologized to former Secretary of the Interior James Watt in 2005.[46] Moyers gave a speech at Harvard Medical School that mentioned a quote that had long been attributed to Watt.

Remember James Watt, President Ronald Reagan's first secretary of the interior? My favorite online environmental journal, the ever-engaging Grist, reminded us recently of how Watt told the U.S. Congress that protecting natural resources was unimportant in light of the imminent return of Jesus Christ. In public testimony he said, "after the last tree is felled, Christ will come back."

Moyers apologized after learning that Watt denied he ever made such a statement and the statement could not be found in the public record. However, in his apology he noted; *You and I differ strongly about your record as Secretary of Interior. I found your policies abysmally at odds with what I understand as a Christian to be our obligation to be stewards of the earth. I found it baffling, when in our conversation of today, you were unaware of how some fundamentalist interpretations of the Bible influence political attitudes toward the environment.*

Since religious fundamentalism can be considered a distorted version of religion, one could argue that the fundamentalist position to which Moyers alluded is just a distorted version of what historian Lynn White, Jr. argued in his seminal 1967 essay entitled "The Historic Roots of Our Environmental Crisis."[47] White argued persuasively that Christianity "not only established a dualism of man and nature but also insisted that it is God's will that man exploit nature for his proper ends." Subsequently, a growing number of theologians and religious studies scholars have attempted a more positive integration of environmentalism and mainstream Christianity. Sallie McFague's work is just one example of this movement and her words help explain why the creationist worldview is problematic for a full appreciation of the natural world: "if the stereotype, even if it is just a stereotype, that we are the dominant creatures and everything is there for our use is connected in any sense with a religious view, then that's part of the problem."[48]

Beyond developing into an absolutely first rate scientist, Charles Darwin was a naturalist who reveled in the glories of nature and recognized the manner in which all life was related. The last paragraph in the first edition of On the Origin of Species represents how he affirmed the sanctity of nature without making any scientific compromises: *Thus, from the war of nature, from famine and death, the most exalted object which we are capable of conceiving, namely, the production of the higher animals, directly follows. There is grandeur in this view of life, with its several powers, having been originally breathed into a few forms or into one; and that, whilst this planet has gone cycling on according to the fixed law of gravity, from so simple a beginning endless forms most beautiful and most wonderful have been, and are being, evolved.[49]*

Let me conclude by relating a personal anecdote. As a scientist, I fully recognize that anecdotes are of limited value when building a scientific argument, but the evolution/creation controversy, particularly the political implications that arise from the rhetoric of the debate, go far beyond science and thus I believe this anecdote may be relevant. In my life, I have had the good fortune to have had three experiences that have literally taken my breath away.

One of those experiences was when I climbed the steep metal spiral staircase from the lower level to the upper level of Sainte-Chapelle on the Ile de la Cité in Paris. Sainte-Chapelle was originally built in 1248 by Louis IX to house two relics he purchased from Byzantine emperor Baldwin II: Christ's crown of thorns and a fragment of the cross upon which he was crucified. The upper level of Sainte-Chapelle is dramatically different from the Gothic architecture that preceded it in houses of worship. Instead, it is one of the first and finest examples of the Rayonnant Style in which glass was incorporated throughout permitting outside light to flood into the chamber. The walls of Sainte-Chapelle are thus comprised of approximately 600 square meters of beautiful stained glass. The glass panes, reading from left to right and top to bottom, in awe-inspiring color, tell the biblical story of humanity. As I rounded the last portion of the spiral staircase and the chapel opened before me, the magnificent colors and images made me gasp.

Another such experience was the first time I jumped from a boat off the coast of Heron Island on Australia's Great Barrier Reef. As I put my head under water and the remarkable diversity of life on the reef came into focus before me, I was again stunned by the beauty I was seeing.

The third experience actually happened twice. Upon emerging into this world, having spent the previous nine months in a protected environment, each of my two sons took his first breath of oxygen – and, at the same moment, I lost mine.

How are these three experiences pertinent to the present discussion? On a very personal level they have demonstrated to me that awareness of the scientific explanations of various phenomena need not detract from the power those phenomena have on a completely different level. Having read a bit about Rayonnant Style architecture before venturing into Sainte-Chapelle, having studied the process of coral reef formation and the nature of reef community structure, and having learned about the process of human childbirth in no way reduced my sense of wonder and joy when I experienced each of those things for myself. Knowledge and awe need not be mutually exclusive. Indeed, there's every reason to believe that they may work synergistically and yield enhanced experiences.

Within each of us, it should be possible to understand and appreciate a wealth of ideas – ideas that are, at times contradictory. Perhaps the biggest challenge we face is to find ways to draw on the best of each, without compromising their essential core, and to increase the level of our knowledge. As we move toward that goal and as we struggle with the cognitive dissonance it invariably entails, we act in a manner that is quintessentially human. And that is something well worth celebrating.

From the Editor....

As Michael Zimmerman points out, a proper sense of the sanctity of nature draws deep meaning from the discoveries of science as related to the origin of nature. Evolution teaches us that nature is millions of years in the making. It also awakens us to the realization that creation is not a static event that happened long ago, but rather, it is an ongoing, unfolding drama of life that is still in action. We are in the moment of creation. God is still creating, and still naming it good. Nature continually finds itself in a holy moment, a moment in which God is still saying "let there be..." The discoveries of science concerning the origins of nature enrich our theological understandings of the value of nature.

Even as evolutionary science leads us to an awareness of the long history of nature's development, it also suggests that humans are a fairly recent part of this story. Our exploration in this study is two-fold. We are exploring the sanctity of nature, and of the human place within it. There is a predominant view that humans are unique in the realm of nature because we have our feet in two worlds. This view teaches that we are a part of nature, in that we have physical bodies, but we are also apart from nature, because we have non-physical souls. Thus, our physicality is a temporary capsule, and the natural world is a temporary stage. In this view, nature has limited value and sanctity. It serves a utilitarian purpose.

We cannot discover a full sense of the sanctity of nature until we wrestle more with a sense of human nature, and of the human place within nature. This is the issue at the heart of our next essay. In the following essay, Nancey Murphy leads us on a journey of rediscovery, as she presents a case for physicalism.

Dr. Nancey Murphy is the Professor of Christian Philosophy at Fuller Seminary in Pasadena, California. She is highly sought as a speaker at national and international conferences on philosophy and the relationship between theology and science. Murphy serves on the board of the Center for Theology and the Natural Sciences, Berkeley, and is a member of the Planning Committee for conferences on science and theology sponsored by the Vatican Observatory.

Chapter 7
Nature and Human Nature:
The Elephant in the Room

by Nancey Murphy

1. Introduction

I'm sure that none of the writers in this book needed to ask what the phrase, "the sanctity of nature," means. Nonetheless, while it is an expression one hears often, it is odd if one thinks about the primary meaning of "sanctity": saintliness, holiness, godliness. These are terms that apply only to humans, and even among ourselves we tend to apply them sparingly! Secondary meanings of "sanctity" include inviolability, being hallowed, or treated with reverence. So to claim that one believes in the sanctity of nature must mean something like "I revere nature"; and it has persuasive overtones, such as "I want you to treat the natural world as inviolable."

But why should we? Are there Christian theological grounds for arguing for the sanctity of nature?[50] I've dabbled in a number of fields, but I've stayed away from ecological ethics because the scriptural grounds have seemed weak and ambiguous. One can cite, for example, God's command to the human to cultivate and care for the Garden, but in the next chapter this same God says "Accursed be the soil because of you." Noah Efron points out that one of many modern Jewish attitudes toward nature is based simply on its being God's handiwork, and, as such, sacred. But this has certainly not been the typical attitude among Christians throughout our history.

There are theological explanations to be found for Christians' disregard of the natural world. Probably the most significant is the Augustinian doctrine of the Fall, which included the disruption of the entire natural cosmos. While most theologians have rejected this cosmic aspect of the doctrine, it is still at work in popular imagination. Contemporary ethologist Frans de Waal has written extensively on the behavior of social animals. An important aim of his work is to counteract a scientific culture that is ready to describe animal behavior in morally negative terms—for example, some chimpanzees are called "cheaters" or "grudgers," and kinship bonds are called "nepotism." Yet these same scientists refuse to use any language with a positive moral tone. De Waal shows that human capacities for morality, such as sharing and caring for the sick or disabled, have quite striking predecessors among certain species of animals. So what explains the preference for viewing animals in a negative moral light? De Waal suggests (in a section titled "Calvinist Sociobiology") that the source is Christian conceptions of the Fall according to which all of nature is corrupted.[51]

I will not pursue the theological issues here, but rather attempt to provide some insight into the way philosophy in the past has led to the denigration of the natural world, and how current developments may help to cultivate adequate respect for our environment today.

Thus, in what follows, I'll first describe the pernicious effects of the philosophical model of the great chain of being. Second, I'll describe the rise and fall of dualist theories of human nature, culminating in a growing consensus among philosophers, scientists, and Christian scholars on the desirability of a physicalist account of human constitution. Finally, I'll argue that a physicalist anthropology, coupled with current knowledge of ecology, provides a sound basis for extending the reach of our theological and ethical concern to the entire ecosphere.

2. The Great Chain of Being

De Waal includes in his defense of animals a critique of what he calls "the age-old half-brute, half-angel view of humanity," a view that he attributes to a convergence of religion, psychoanalytical and evolutionary thought.[52] I take its appearance in these more recent sources to be secondary to the idea famously described by Arthur Lovejoy as "the great chain of being."[53] This is a Hellenistic idea that shaped Western consciousness from the days of classical Greece through the end of the Middle Ages. According to the Christian version, everything that exists, from rocks to God, can be arranged in a hierarchy: inorganic materials, plants, animals, humans, angels, and God. The great ontological divide here is not between Creator and creatures, as I think it should be for readers of the Bible, but rather between matter and spirit ('spirit' understood in gnostic rather than Pauline terms). Humans on this view are "amphibious" creatures: their bodies are on the lower side of the great divide, their souls above. This being a hierarchy of value and not merely a classificatory scheme, Westerners have grown accustomed of thinking of themselves as distinctly superior to animals in moral terms. I believe that this is one of the sources of negative attitudes toward animals. They are "beastly," while we are (when we behave in ways commensurate with our place in the hierarchy) "humane."[54]

The holdover from this very old worldview, I believe, is part of the explanation for why there has been resistance to accepting the fact of our close kinship with animals, as well as for attitudes permissive of the exploitation of the rest of nature.

The "amphibious" view of humans was dependent, of course, on dualist anthropology, to which I turn in the next section.

3. The Rise and Fall of Anthropological Dualism

I've subtitled my essay "The Elephant in the Room" because of my perception that strident ethical disagreements are often based on conflicting views of human nature, and yet these views are never made explicit. Why do opponents of abortion and stem-cell research argue that both involve the taking of human life? Because they are dualists and believe that the soul is present from the moment of fertilization. When Dolly the sheep was cloned I received a call from a reporter who seemed frustrated that I had no strong condemnation of the idea of cloning humans. After his repeated attempts to provoke me to express some sort of horror at the prospect, light dawned for me. I asked him: "Are you imagining that if we try to clone a human being we'll clone a body but it won't have a soul? It will be like the zombies in science fiction?" "Yes, something like that." "Well," I said, "Don't worry. None of us has a soul and we all get along perfectly well!"

Dualism may be playing a similar role in some Christians' lack of concern for the natural world. While Christian teaching at its best has denied that the human soul is divine, theologians have always been working against the assumption that souls are bits of "Godstuff."[55] This tendency has made it all too easy to identify that which is distinctively human with God, the sacred one, and to contrast it with all of nature.

I speak often on the topic of human nature, and because of the remarkable silence on the dualism-physicalism issue, I've had to resort to informal polling of my audiences to find out what they believe. I've found that among laypeople (neither theologians nor scientists) the vast majority are either dualists or trichotomists (body, soul, and spirit). If I am correct that dualism supports a disregard of nature, then it is encouraging that discussions of these issues are now beginning to take place, and that the direction of change is toward physicalism.

3.1 A Short History of Dualism

When I first began to study the place of dualism in Christian theology I was disappointed to find no comprehensive history of the relevant issues. The relevant terms, "soul," "immortality," "resurrection," did not even appear in the indices of major histories of doctrine. I therefore resorted to patching together my own account of the history by consulting reference works from various periods.[56] Here is my account, which is certain to be controversial due to its brevity.

First, there is no concept of the soul anything like that of today's Christians in the ancient Hebrew Scriptures. This conclusion has been widely agreed upon for half a century. Also there was no clear concept of life after death until close to Jesus' day, at which time some Jews expected bodily resurrection at the end of history and some others had adopted body-soul dualism and the belief in the immortality of the soul. However, New Testament authors largely adhered to the ancient Hebraic concept of the person. Thus, they were closer to contemporary physicalism than contemporary body-soul dualism. [57]

Post-biblical teaching shows the gradual development of the dualism that remained central to Christianity for centuries. Early teaching on the afterlife, such as that of Clement of Rome (writing approx. 95 CE), focused on immortality as a gift from God, and as a consequence of resurrection of the body, with no mention of a soul. The fate of those who were not saved was simply death. The first mention in Christian teaching of an immortal soul was in the Epistle to Diognetus (approx. 130). Athenagoras was the first to link a philosophical conviction of the natural immortality of the soul with a Christian doctrine of the punishment of the wicked, and to conclude that the damned would suffer eternally. By the time of Augustine

(354-430) the doctrines of body-soul dualism and immortality of the soul were firmly entrenched in Christian teaching. From Augustine's day until that of Thomas Aquinas (1225-74), Christian dualism was based on Platonic philosophy. Thomas developed a moderate Aristotelian dualism, according to which the soul is the form of the body. This remained an influential account of human nature through the Renaissance, and is still the official Catholic position. Meanwhile, during the Reformation, Protestants tended to return to Augustinian theology with its Platonic account of soul and body.

3.2 The Rise of Physicalism

Christian scholarship, philosophy, and science have all played a role in the discrediting of dualism and the promotion of a physicalist anthropology. The first scientific contribution was the rise of modern physics. The demise of Ptolemaic astronomy also spelled the end of Aristotelian physics. When the latter was replaced by atomism, the Aristotelian-Thomist account of the soul as the form of the body no longer made sense in metaphysical terms—in the new modern worldview there simply are no such things as forms. Two possibilities for modern replacements are represented by René Descartes's radical dualism (again influenced by Augustine) and Thomas Hobbes's reductive physicalism. Apart from an Idealist interlude in the nineteenth century these have been the predominant philosophical competitors until the late twentieth century.

The theory of evolution has had wide-ranging effects on human self-understanding, but relates to the dualism/physicalism debate in that it raised, for some, the question of why humans should be thought to have souls if their close animal kin do not. Others responded with an emphasis on dualism as the very thing that distinguishes us from animals. The third major scientific impact is taking place right now due to the influences of contemporary neuroscience. It is becoming increasingly obvious that the functions once attributed to the soul or mind are better understood as functions of the brain. These developments in neuroscience, along with the judgment that no account can be given of mind-body interaction, have resulted in a near total rejection of dualism in philosophy of mind. Current debates in philosophy focus on the issue of reductionism—is there any way to argue for the causal efficacy of the mental, or is it really brain functions that determine human thought and behavior. I return to this issue below.[58]

While it is fair to say that developments in neuroscience provoked the shift to physicalism for many philosophers, it would be a mistake to say the same for Christian scholars. While the science has brought new attention to the issue, the rejection of dualism began a century ago in biblical studies. In 1911 H. Wheeler Robinson argued in The Christian Doctrine of Man that the Hebrew idea of personality is that of an animated body, not that of an incarnated soul.[59] From that point on a variety of scholarly developments favored the rejection of dualism as the position of the Bible. One was the rise of neo-orthodox theology after World War I. Karl Barth and others made a sharp distinction between Hebraic and Hellenistic conceptions, and strongly favored the former. Barth also argued for the centrality of the resurrection (rather than immortality) in Christian teaching. The biblical theology movement in the mid-twentieth century continued to press for the restoration of earlier, Hebraic understandings of Christianity. A decisive contribution was Rudolf Bultmann's claim in his Theology of the New Testament that Paul uses soma ("body") to characterize the human person as a whole.[60] In 1955 Oscar Cullmann gave the lectures that were published as "Immortality of the Soul or Resurrection of the Dead: The Witness of the New Testament." Here he drew out the contrast between biblical attitudes toward death, along with expectation of bodily resurrection, and Socrates' attitude given his expectation that his soul would survive the death of his body.[61]

Thus, in the minds of many, the issue was settled in favor of physicalism as original Christian teaching. Of course, this account is incomplete, in that many conservative Protestants either ignored or attempted to refute these developments; the issue is only now receiving the attention among Evangelicals that it received from liberals fifty years ago. Catholics are divided as well: biblical scholars tend to be physicalists, while some theologians subscribe to the Thomist theory. In any event, it is clear from my informal surveys that however thoroughly dualism has been rejected in the academy, this information did not reach the Christians in the pews.

Now, if it is the case that body-soul dualism is foreign to the Bible, how is it that Christians for centuries could have been so wrong in believing dualism to be biblical teaching? A crucial distinction comes from New Testament scholar James Dunn. Dunn distinguishes what he calls "aspective" and "partitive" accounts of human nature. Greek philosophers tended to be interested in a partitive account: what are the essential parts that make up a human being? In contrast, the biblical authors were interested in an aspective account. Here each 'part' ("part" in scare quotes) stands for the whole person thought of from a certain angle.[62]   So, for example, Paul's distinction between spirit and flesh is not the later distinction between soul and body. Paul is concerned with two ways of living: one in conformity with the Spirit of God, and the other in conformity to the old aeon before Christ.

Dunn's insight explains how Christians for hundreds of years could have taken dualism to be scriptural teaching. The Old Testament was translated into Greek (the Septuagint). Both the Old and New Testaments then contained the Greek terms that in the minds of philosophers referred to constituent parts of humans, and Christians have obligingly read them and translated them in this way for centuries. The clearest instance is the Hebrew word nephesh, which was translated as psyche in the Septuagint and later translated into English as 'soul.' More recent translations use a variety of English words. For example, Genesis 2:7 used to read: ". . . The Lord God formed man of the dust of the ground and breathed into his nostrils the breath of life and man became a living soul." Recent translations say that man became a living being, (NIV) or a living creature (REB).

So I conclude that there is no such thing as the biblical view of human nature insofar as we are interested in a partitive account. The biblical authors, especially New Testament authors, wrote within the context of a wide variety of views, probably as diverse as in our own day, but did not take a clear stand on one theory or another. What the New Testament authors do attest is, first, that humans are psychophysical unities; second, that Christian hope for eternal life is staked on bodily resurrection, not an immortal soul; and, third, that humans are to be understood in terms of their relationships— relationships to the community of believers and especially to God.

3.3 Nonreductive Physicalism

I mentioned above that current debates in philosophy of mind focus on the issue of reductionism, which can be represented by means of the following question: If it is the case that the brain performs the functions once attributed to the mind or soul, then how can it fail to be the case that all human thought and behavior are merely the result of blind neurobiological processes? Addressing this problem requires making explicit the powerful role that causal reductionism has played throughout the modern worldview. The assumption, taken over from Epicurean atomism at the beginning of modern physics has been that in any complex entity or system, it is the behavior of the parts that governs the behavior of the whole. The (mechanical) clock served as a powerful metaphor reinforcing this assumption.

At present there is a change in the understanding of complexity that some describe as a Kuhnian paradigm change across all of the sciences.[63] Others go so far as to say that the development of "complex adaptive systems theory" provides the basis for an entirely new worldview.[64] Over fifty years ago, philosophical theologian Austin Farrer was groping for language to describe the change that is just now becoming increasingly evident. He distinguished between two types of systems. The familiar type is one in which the pattern of the whole is a simple product of the behavior of its parts. The other sort of system is one in which "the constituents are caught, and as it were bewitched, by larger patterns of action." As examples he cites the molecular constituents of cells and cells themselves within the animal body. Furthermore, "[new] principles of action come into play at successive levels of organization." Farrer recognizes that he is denying deep-seated reductionist assumptions, but maintains that "the intransigence of the [reductionistic] physicists . . . need not contradict the claims of the biologists to be studying a pattern of action which does real work at its own level."[65]

These "other sorts" of systems are now widely recognized: Even though the systems are composites of basic physical constituents, the causal powers of such a system are not determined solely by the physical properties of its constituents and the laws of physics. They are also determined by the organization of those constituents within the composite. These patterns have "downward causal efficacy" in that they can affect which causal powers of their constituents are activated. Thus the whole is not any simple function of its parts, since the whole at least partially determines what contributions are made by its parts. Such patterns or entities are stable features of the world, often in spite of variations or exchanges in their underlying physical constituents. Many such patterns are self-sustaining or self-reproducing in the face of perturbing physical forces that might degrade or destroy them (e.g. DNA patterns). Finally, the selective activation of the causal powers of such a pattern's parts may in many cases contribute to the maintenance and preservation of the pattern itself. Taken together, these points illustrate that higher-order patterns can have a degree of independence from their underlying physical realizations, but without altering the underlying laws of physics.[66]

The human brain is the single most complex entity we know of. When we consider also that the brains of individuals are enmeshed in immensely complex relationships, influenced by various aspects of culture, it makes much more sense to think of cultural contexts and individual behavior having a reciprocal effect on the brain, rather than assuming that the brain unilaterally determines human thought and behavior.

Much more needs to be said to explain the means by which the "downward efficacy" of the system results in behavioral flexibility, free will, and moral responsibility, but clearly it is this new set of "systems" concepts that will provide the resources.[67]

4 Human Nature and Nature

I have argued so far that nonreductive physicalism is a theologically, scientifically, and philosophically sound theory of human nature, and am now at the point of being able to apply it to the topic of the moral and theological status of nature. There are two features of physicalism that make it particularly relevant. First, scientific knowledge is making it increasingly obvious that humans cannot be understood apart from their embeddedness in nature. Quite interesting developments in cosmology have led some scientists to conclude that we live in an "anthropic" universe. Even more important for theological ethics are the developments in ecology, showing that we are not only physical beings, but "eco-physical beings." [68] A second feature is theological. I've noted in passing that while immortality is the expectation for life after death associated with dualism, physicalism calls for eternal life conceived in terms of bodily resurrection. When expectation for the resurrection of human bodies is considered within the context of our eco-physicality, the implication is expectation for cosmic transformation, of which the resurrection of Jesus is a foretaste. And if the entire cosmos merits transformation and eternal preservation in the eyes of God, does this not argue for its "sanctity"?

4.1 Humans as Eco-Physical Beings

I've already noted the change in translation of Genesis 2:7; in recent translations the world "soul" has disappeared. What even the new translations fail to show is a literary device in the Hebrew that adds emphasis to the materiality of the human. The term "man" translates adham, which is not a man's name but a generic term for humans. The word for "ground" is adamah. So the pun, "adham formed from adamah" is a literary device that highlights the dusty origin of the species. We can recapture the imagery in English by describing ourselves as humans made from humus. This dust of the ground we now know to be, ultimately, star dust. The heavy elements form in stars and are distributed when the stars explode—to become planets and plants and people. In cosmology there is a fascinating discussion of the so-called anthropic issue. This is based on calculations showing that very small changes in any of the numbers that go into the basic laws of physics would have resulted in a universe in which no life is possible. For instance, if the strength of gravity been slightly higher our universe would have collapsed in on itself too quickly for stars, planets, and life to evolve. These anthropic calculations have led some interpreters to describe the universe as fine-tuned for life. Some see it as the work of God; others dismiss it as mere chance. But it has led a number of thoughtful scientists to raise questions about the significance of human life in the universe. "I do not feel like an alien in the universe," says physicist Freeman Dyson.[69] Indeed, this world is our home. The vastness of the universe was once taken to speak to human insignificance. But in light of the anthropic calculations we can say that the universe needed to be as immense as it is, and as old as it is, and as full of stars as it is, in order for us to be here.

Even more important than current appreciation of our place in the cosmos as a whole is the vast amount of knowledge that constitutes the relatively new science of ecology. John Mustol argues more persuasively than I can summarize in this short piece that ecological ethics must become central to Christian ethics.[70]   We are a part of the Earth's ecology, the laws of ecology are as significant to our survival and that of our biosphere as are the laws of physics. They must be taken into account as we strive to live in closer accord with them. Mustol states:

As physical organisms, we humans are subject to all the patterns, principles, limits, and "house rules" that govern the existence of all organisms living with us on earth. We must work within the "interdependencies, resources, and constraints" imposed upon us by the biosphere and all its systems. We are bound up within a system of "checks and balances, controls and feedback loops" that we must learn about and obey. Even in this day of our enormous technical and energetic power, these ecological realities circumscribe our existence.[71]

The "house rules" that Mustol lists include the recycling of finite resources, the facts of energy flow and exchange, the webs of food chains, population dynamics, and the concepts of carrying capacity and ecological footprint.

Mustol argues that adequate attention to these facts of life will require a significant change in human consciousness. We have evolved to focus our attention on groups of our own species; to extend that focus to our ecological setting requires a significant shift in self-perception. I have already introduced his term "eco-physical humans." A second terminological shift is to abandon the distinction between ourselves and "our environment" since it implies that we humans are somehow at the center and that all else is what surrounds us. We are a part of an ecology. As James McClendon wrote, we cannot properly do Christian ethics if we do not take with full seriousness the fact that we are "organic constituents of the crust of the planet." [72]

4.2 Ecology and Eschatology

When I speak about physicalist anthropology to dualists, they often hear it as bad news regarding what happens at death, since their hope for the long-term future is based on the survival of an immortal soul. However, I believe that physicalism sounds a more hopeful note. Physicalism and dualism are each bound up with an entire worldview, and I see the worldview of the physicalist as bearing great promise for the future of our species and our planet.

Dualism belongs to a worldview that owes a great deal to Plato. The title of a chapter on Plato's philosophy aptly characterizes his central contribution: "This World Is Not our Home."[73] Plato invented the notion of a nonmaterial realm transcending this corruptible material world. The dualist view of the person mirrored this cosmic dualism. The human soul, immortal, belongs to the transcendent realm of the Forms (or Ideas), and life in the body is temporary imprisonment. Value resides in the other world; in fact, some of Plato's followers counted matter as essentially evil.

The Western imagination has been formed by a storyline that incorporated into Christian teaching Plato's otherworldliness and the ubiquitous ancient idea of a Golden Age in the past, followed by a catastrophic Fall. Human misery and all natural evil and imperfection are attributed in one way or another to this Fall. While original Christian teaching about the end of the story centered on bodily resurrection, later Christians have focused more on a Platonic hope for the soul's escape, to live forever in a transcendent Heaven.

It is a mistake, of course, to think that evolutionary biology denies divine creation. However, evolution, along with current theories about the history of the universe as a whole, does contradict this ancient storyline from Original Perfection to Fall to Restoration. Science provides the backbone for a new storyline, and calls for a strikingly different worldview. The new story sees no perfection in the past, at least not in terms of human values. Rather, the universe immediately after the Big-Bang was chaotic and composed of the simplest of ingredients—for some time, composed merely of gasses.

Neither cosmology nor biology alone can provide criteria for assessing progress—for example, biology cannot say why it is better to have mammals and humans than a lot more insects. But from our human point of view we cannot but say that the evolutionary process involves progress—in particular, progress to the point where its products can raise questions of value! There are longstanding disagreements over the question of whether evolution provides grounds for optimism about the future—that is, it may or may not entail progress in human history. But we can say at least that the new story frees us from the sense of tragic loss associated with the traditional story and its catastrophic Fall.

In contrast to the Platonic worldview, as noted above, the sciences tell us that this world is very much our home—not just planet Earth, but the whole universe. The traditional storyline attributed toil, suffering, and death, human and animal alike, to the Fall. Much suffering, of course, is the direct result of ubiquitous human sin. But in this new worldview we must raise questions afresh: Why are we so bent on sinning, if not because of a disorder in our souls caused by Adam's defection? And why are there natural disasters, if not because of the fall of the angels?

The old storyline says that animal nature was corrupted by human sin. But here is an interesting twist on the relationship between animals and human morality: Again I turn to de Waal's analysis. He speculates that sharing and other positive social behaviors first evolved among animals that needed to form packs for hunting and killing prey. "If carnivory was indeed the catalyst for the evolution of sharing, it is hard to escape the conclusion that human morality is steeped in animal blood. When we ... ship food to starving people, or vote for measures that benefit the poor, we follow impulses shaped since the time our ancestors began to cluster around meat possessors." [74] An ironic reversal of the traditional view that carnivory only began with human sin.

A deeper explanation for the toils and trials of human and animal life is found in physics. The second law of thermodynamics determines that any system will degrade, decay, run down, if not supplied with energy from outside. This law is intimately involved in all biological processes—it is part of the fine-tuning that makes life possible—but it is the ultimate cause of hunger, of the need for shelter, for hard work, and, ultimately, death. The surface of the Earth itself needs constant replenishment—so earthquakes and volcanoes are necessary for life.

In sum, the character of the natural world is not the product of sin, human or angelic, but, through a theist's eyes, it can be seen to have been precisely designed for creatures like us. The suffering from nature and within nature can be seen as unwanted yet necessary by-products of the very laws of nature that make our existence possible.

The old storyline from Creation to Fall to Restoration must yield to a storyline from creation, through slow and painful development, from the simple and chaotic to the complex and orderly. But then what? What is the final end of this process? Has it a goal or purpose? Here science provides no help and we must turn specifically to religion. How, in this new context do we retell the story of God's purposes and plans for the universe and the human race?

Here you must allow me to be a bit speculative. I believe we can tell the story of the universe as a story of God creating something that was not God, something at first formless and chaotic—as unlike God as anything could be. God's creative hand has gradually brought order out of chaos and increasing complexity out of lower forms of order. But to what purpose? So that, ultimately, from that which is non-God there would
emerge beings who are images of God, mirrors of the divine likeness—dust-creatures with the capacity to know and appreciate the universe (this includes doing science), and especially to know and respond in love to the Creator. I appreciate the image suggested by my Jesuit friend Ed Oakes: "...*humans are the priests of the universe. We serve as mediators between God and the rest of creation, for we alone can hear God's word and consciously carry out God's will. As such we mediate God's will to the rest of creation. We alone have tongues to praise God on behalf of the rest of the universe.*[75]

Proper Christian hope for the future is based not on "soul-ectomy," the surgical removal of the immortal soul, but rather on resurrection.[76] We say "resurrection of the body" but we should say resurrection of the person, the whole person. And the only vision of the end of the world that is fully consistent with the hope of resurrection is a transformation of the whole cosmos, a transformation of which the resurrection of Jesus on Easter is first fruits. We can say nothing of what this transformation will be like in scientific terms because all science is based on the way things are in this aeon. But we can say much about that new world in moral terms. This will be a world whose character Isaiah evoked in his prophecy:

For I am about to create new heavens and a new earth....
I am about to create Jerusalem as a joy, and its people as a delight. ...
No more shall the sound of weeping be heard ... or the cry of distress.
They shall build houses and inhabit them;
they shall plant vineyards and eat their fruit. ...
Before they call I will answer,
while they are not yet speaking I will hear.
The wolf and the lamb shall feed together,
the lion shall eat straw like the ox,
They shall not hurt or destroy on all my holy mountain, says the Lord.
(Is. 65:17-25)

Note that this is a social vision—the re-creation of city life. It is a vision of unimpaired, immediate relation to God. And it is a vision of a whole new cosmos—new heavens as well as new earth—in which humankind and all of nature will be reconciled.

Jesus' teaching about the world to come focused more narrowly on human reconciliation and reconciliation with God. A common image for the Kingdom of God is a wedding feast to which all are invited to share the Bridegroom's bounty. But the Apostle Paul notes that the whole creation waits in eager longing to be set free from its bondage to decay (Rom. 8:19-23).

If we reject the Neoplatonic vision of the "flight of the alone to the Alone," and return to the biblical view of the Rule of God "on earth as it is in heaven," we find a vision of end of time that shows the ultimate value of sociality; that shows that history is meaningful, for past achievements are not left behind but transformed, past sorrows add poignancy to present joy. Finally, it is a vision that shows there to be ultimate value in our care for and harmony with the whole of nature.

BIBLIOGRAPHY

Bouma-Prediger, Steven, For the Beauty of the Earth: A Christian Vision for Creation Care. Grand Rapids, Baker, 2001.

Green, Joel B., Body, Soul, and Human Life:The Nature of Humanity in the Bible. Grand Rapids, Baker, 2008

Krebs, Charles J. Ecology: The Experimental Analysis of Distribution and Abundance. San Francisco: Benjamin Cummings, 2001,

Murphy, Nancey. Bodies and Souls, or spirited Bodies? Cambridge University, 2006.

From the Editor...

In the previous essay, Nancey Murphy confronts the dualistic path that has led Christianity away from its root ideas concerning humanity's place in and relationship to nature. Nature is not a temporary setting. It is our home. We are very much a part of this physical universe. As such, we have a deep, symbiotic relationship to nature, and to all the other creatures that inhabit it. Yet, the discoveries of science in the field of evolution have taught us that nature has existed for millions of years before modern humans arrived on the scene. Nature has hosted scores of life forms that have come and gone before humans arrived. As Jennifer Wiseman illustrated in our opening essay, nature includes numerous star systems and galaxies of which our star system is a very small part. All of this sheds light on our notions of the sanctity of nature, but it also begins to call into question some of the models we have employed to describe our human relationship to nature. In our next set of essays, we will explore some of the theological concepts that move us toward a new sense of an appropriate theological anthropology. Whenever we entertain a theological study of what it means to be human from a Christian point of view, it is appropriate for us to begin with the life of Jesus Christ. This is the direction that our next essay will take us, as Rebecca Flietstra addresses the doctrine of the Incarnation as it speaks to our theme.

Dr. Rebecca J. Flietstra is Professor of Biology at Point Loma Nazarene University in San Diego. She earned a BA in biology from Calvin College and her PhD in physiology from Kansas University Medical Center. Rebecca won a Templeton Science and Religion Course Award in 1999 and gave science and religion seminars at Oxford in 1999–2001.

Chapter 8
A Creation Redeemed by Incarnation

by Rebecca J. Flietstra

*In the beginning when God created the heavens and the earth,
the earth was a formless void and darkness covered the face of the
deep, while a wind from God swept over the face of the waters. Then
God said, "Let there be light"; and there was light. And God saw that
the light was good; and God separated the light from the darkness.
God called the light Day, and the darkness he called Night. And there
was evening and there was morning, the first day.*
Genesis 1: 1-5

*In the beginning was the Word, and the Word was with God,
and the Word was God. He was in the beginning with God. All things
came into being through him, and without him not one thing came
into being. What has come into being in him was life, and the life was
the light of all people. The light shines in the darkness, and the
darkness did not overcome it. ... And the Word became flesh and lived
among us, and we have seen his glory, the glory as of a father's only
son, full of grace and truth.*
John 1: 1-5, 14

The well-known prologue in the gospel of John (John 1: 1-18)
both echoes and reframes the creation story of Genesis one. Both, of
course, open with "In the beginning..." and discuss the origin of all
created things and of all life. Yet the details are markedly different. In
Genesis one, the Creation rather quickly comes into being as God
speaks. God's speech serves to create, to separate ("light from the
darkness," "waters from the waters"), and to repeatedly declare that
all Creation is "good." Such declarations are not found in the
Johanine retelling. While Creation is still from God (and thus
presumably good), it also has become distant and unable to recognize
the Creator. Thus we read that "the world came into being through
him; yet the world did not know him" (v. 10b). God's Word thus does
not enter Creation in order to impose separations, but in order to bring
together: to restore a right relationship between human and divine by

becoming human and divine; to reconcile all of Creation with God by becoming created and Creator.

The Incarnation (God's taking on human nature in Jesus) stands at the center of Christian faith, theology, confessions and practice. Being central, of course, does not mean that the Incarnation has been easy to articulate or to understand. Indeed, the early Church Fathers frequently wrestled with this doctrine, carefully co-opting philosophical frameworks from the surrounding cultures and, at the same time, creating new terms and analogies as they sought to place boundaries between orthodoxy and heresy. These conversations and debates have not only shaped how the Church has understood the nature of Jesus Christ, but have also forced theologians to articulate Christian beliefs about God, about humanity, and about Creation. For if Jesus is fully divine, we must know what we mean by "God." And, if Jesus fully participates in our humanity, then that must tell us something about who we are and what we should be—not merely on a metaphysical plane, but also as bodies, as biological entities within a physical Creation.

In this essay I will first briefly summarize Christian beliefs about the Incarnation. Secondly, I will discuss how the doctrine of Incarnation has shaped Christian beliefs about ourselves, i.e., about what it means to be human. Finally, I will discuss how this doctrine might inform Christian beliefs about the entire Creation.

The Human Jesus

Several Christmases ago, when my oldest nephew was five, my sister talked with him about Jesus' birth—about the angels and the shepherds, the magi and the star, and about Joseph and Mary with baby Jesus in the stable. She marveled at how this was all a great mystery, that we really can't understand how God became a human baby and lived here on earth. My nephew disagreed: "No, mom, the real mystery is how Jesus gets born every year. I know he's going to be born this Christmas, and I remember him being born last Christmas, and I think he was born in the years before that. That's the real mystery."

It is indeed a mystery. In their celebration of Christ's birth and their commemoration of Jesus' death and resurrection, Christians often do appear to forget that between birth and death were three decades in which Jesus lived as an infant, a child, an adolescent, and an adult. The very earliest Christians, of course, didn't forget this. Some had known Jesus as a child, or had met Jesus' mother and siblings. They knew that Jesus had been circumcised, had studied the Torah, worshipped at synagogues, prayed, sacrificed, and participated in pilgrimages and festivals. Jesus experienced times of hunger and thirst, slept and partied, gave sermons and told stories, cried at the tomb of a good friend, and even bled and died. After Jesus' death, resurrection and ascension, and as the church and its doctrines began solidify, Christians began to reflect on the meaning of Jesus' life. They recognized that the end of Jesus' life could not be understood apart from his three decades as a Jewish man. For Jesus' salvific role required a full identification with humanity as he experienced all the various aspects of the human condition.[77] That is, only by truly "dwelling among us" as a fellow human and not as an alien visitor or dabbler, could Christ represent and redeem the rest of humanity. Thus Cyril of Alexandria insisted,

When the wise evangelist introduces the Word as having been made flesh he shows him economically, allowing his own flesh to obey the laws of its own nature. It belongs to manhood to advance in stature and wisdom, and one might say in grace also, for understanding unfolds in a certain fashion in each person according to the limits of the body. It is one thing in infants, something else in grown children, and something different again for adults. It would not have been impossible, or impractical, for God the Word who issued from the Father to have made that body which he united with himself rise up even from its swaddling bands, and bring it straight to the stature of perfect maturity. One might even say that it would have been plain sailing, quite easy for him to have displayed a prodigal wisdom in its infancy; but such a thing would have smacked of wonder-working, and would have been out of key with the plan of the economy. No, the mystery was accomplished quietly, and for this reason (that is economically) he allowed the limitations of the manhood to have dominion over himself. This was so arranged as part of his "likeness to us," for we advance to greater things little by little as the occasion calls us to assume a greater stature and a concomitant mentality.[78]

In other words, even though God could possibly have chosen to create a full-grown adam—bypassing all the bruises and risks of childhood, the vulnerabilities of family and friends, and the finiteness of being human—God instead opted to become fully and truly human.

The Incarnate Jesus

This, of course, is quite a claim. It's easy to recognize that Jesus was human; it's much more challenging to simultaneously declare that he was and is divine. Yet shortly after Jesus' death and resurrection, the church recognized and worshipped the human Jesus as God.[79] For the earliest Christians—for Jesus' disciples, for the apostle Paul, and for the earliest converts—the belief that a human being could also be the one God was (as it is for us) an amazing leap. Certainly many of the surrounding cultures of that time, including the dominant Roman culture, believed that the gods and humans (generally attractive young women) intermixed and propagated, but Christianity's roots were within Judaism. For centuries the Jews had held and proclaimed the Shema: "Hear, O Israel: The LORD our God, the LORD is one" (Deut. 6:4). Thus first century Jews (as well as many Jews today) would have understood the worship of Jesus as blasphemy and idolatry. Yet these early, predominantly Jewish, Christians not only recognized Jesus as fully God and fully human, but also claimed that the Messiah must necessarily be incarnate God. As St. Athanasius wrote on several occasions, the "[Son] has become Man, that He might deify us in Himself." [80]

Interestingly, as the early Christians struggled to understand and to articulate this doctrine (and the related doctrine of the Trinity) many began to emphasize Jesus' divinity while rejecting his humanity. No longer contemporaries of Jesus, some carefully distinguished between the human aspect (Jesus) and divine aspect (Christ) of the God-man, where the divine soul had become trapped in the frail—and even evil—human body. Others portrayed Jesus' body as an illusion or, at the most, a shell used by God to act as if he had become human. Many denied the real birth and/or death of Jesus' body, claiming that the Christ hypnotized the crowd into believing his body was on the cross, when in reality Simon Magus or Judas suffered through the crucifixion. All these solutions reflected a belief that human nature—indeed, all the physical world—was irredeemably tainted by sin and evil, such that redemption required an escape into a purely spiritual realm.

The temptation to protect Jesus' deity by denying his humanity still exists today. For example, Young Earth Creationist Henry Morris frequently claimed that the Virgin Mary had no genetic connection with the child incubating in her womb. As articulated in his Defender's Study Bible: *"That holy thing" was placed directly in Mary's womb by God "the Holy Ghost" (Luke 1:35) and thus was uniquely "the seed of the woman" (Genesis 3:15). Just as the body of "the first Adam" was directly formed by God (Genesis 2:7), with no genetic connection to either father or mother, so the body of "the second Adam" (I Corinthians 15:45) was directly formed by God (Hebrews 10:5) with no genetic connection to either parent. Since the very ground was brought under God's curse because of sin (Genesis 3:17), all the elements of the ground ("the dust of the earth") out of which the bodies of Adam and all their descendants had been formed were contaminated with the "bondage of corruption" (or decay— Romans 8:21, 22). This was just as true of Mary's body as of Joseph's, so there could have been no natural genetic connection of Jesus' body to Mary's, any more than to Joseph's. The "holy thing" placed in Mary's womb by the Holy Spirit could have been nothing less than a special creation, just as was the body of Adam. Otherwise, like all men born of women, Jesus would have inherited both physical defects and the sin-nature of Adam and Eve. This could only have been prevented by a miraculous cleansing of the conceptus, and this, of course, would be a special creation. Jesus was the only begotten Son of God, as well as the son of Mary, but He was not the Son of God and Mary.[81]*

This claim of Morris' sharply contrasts with the writings of Irenaeus (2nd century A.D.) who, rather than focusing on the presence or absence of parents, observed that Adam was formed from the earth and, therefore, the Son must also be formed from the physical world. For, if the Son had not come from "the hand and workmanship of God" Irenaeus considered that Christ would have been "an inconsistent piece of work." Moreover, "if He did not receive the substance of flesh from a human being, He neither was made man nor the Son of man; and if He was not made what we were, He did no great thing in what He suffered and endured."[82] Gregory of Nazanzius (4th century A.D.) even more acutely observed, "The unassumed is the unhealed, but what is united with God is also being saved."[83] In other words, if Jesus wasn't truly human—truly incarnate—than humanity has no hope of redemption and salvation.

The Created Jesus

Full humanity includes not only growth from a fertilized egg into adulthood, but also encompasses a position within the broader world. For biologists, humans have been classified as Homo sapiens, a species in the order Primates, of the class Mammalia, of the Chordate phylum, of the Animal kingdom. The human genome consists of some 20,000-25,000 genes[84] that are distributed among forty-six chromosomes, and that encode for a currently inestimable number of proteins—at least several orders more than the number of genes.[85] Human females have a cryptic ovulation and can be sexually active throughout their menstrual cycle; conception is followed by a nine-month gestational period. Unlike the other large mammals that tend to have a single offspring per gestational period, human infants are completely helpless at birth and highly dependent on adult care for the first several years of life. Humans walk upright on two legs (bipedal) and are mostly hairless. We have opposable thumbs and can develop and use tools. We have large brains, use language, have self awareness, participate in cultures, worship and believe, and even anticipate our own deaths.

This combination of traits makes the human species unique, yet also connects us with the rest of creation. Certainly none of our atoms are unique. The most common molecule in our body, water, is also the most common molecule found in other living organisms. Our bodies are dynamic systems, taking in substances in order to build and rebuild structures, while tearing apart and metabolizing other molecules. Each human body has a measure of continuity, but at the microscopic and atomic levels each body is also changing on a daily basis. Every day the body of an adult human tears apart and rebuilds 2-3% of its proteins. The amino acids we use to build these proteins are not just found in us, but are present in other plants and animals. The genetic information we store in our chromosomes may have a unique pattern, but it still is incredibly similar to other species.

Several studies, for example, indicate that the sequence of nucleotides (the order of the DNA building blocks) in humans has a 98.5% similarity with chimpanzees.

For biologists, this commonality is no coincidence, but reflects a common ancestry. Indeed, while chimpanzees and bonobos may be Homo sapiens' closest living relatives, we recognize that all living organisms are related to humans via a common ancestry. Paleontologists currently estimate that the first cells appeared in the fossil record 3.5—4 billion years ago, shortly after the molten earth had sufficiently cooled. The earliest cells did not have internal structures that separated their various internal functions. Those divisions only arose an estimated 2.1 billion years ago, as some cells began to use membranes to produce organelles (specialized structures that carry out specific functions). A billion years later, some descendants of these eukaryotic cells began to form simple multicellular organisms. During the Cambrian period (543—510 million years ago), more complex body forms arose, including the Chordate body form with its characteristic hollow notochord under its dorsal surface. In vertebrates the notochord disappears after instigating the development of the spinal cord and brain. Shortly after the Cambrian period plants, fungi, and animals started colonizing land. The first mammals appeared in the early Mesozoic era (245—206 million years ago) and were probably small, shrew-like creatures in a landscape dominated by dinosaurs. As the dinosaurs became extinct mammals filled the now-vacant niches until they dominated the modern Cenozoic era (from 65 million years ago to the present). Primates first appeared 60 million years ago; 30 million years later the ancestors of modern apes and monkeys evolved. Our hominid ancestors appear to have separated from the ancestors of the other great apes (chimpanzees, bonobos, gorillas, and orangutans) more than 5 million years ago. The earliest named species of the Homo genus, H. habilis, lived 1.9 million years ago. Finally, H. sapiens itself appears to be about 150,000 years old.

Our biological history is important not simply because it recounts our ancestry, but also because it recounts Jesus' biological ancestry. When God the Son took on human flesh, he took on our entire evolutionary history and our daily interdependence with the rest of creation. The Son of God's willingness to participate in the physics, chemical reactions, and biological processes of the physical world reveals a God intimately involved in the realm of creation. Just as we (H. sapiens) share a common ancestor, a shared genetic code, and a shared biochemistry with all other species here on earth[86], so Christ now shares in the world's evolutionary history and ecological web of life.

In this light, Gregory of Nazanzius' observation—"The unassumed is the unhealed, but what is united with God is also being saved"[87] —now acquires an even broader meaning.

The divine condescension that accepted the form of a slave and death on a cross accepted also our bestial ancestry. "That which is not taken is not healed,"' and by his incarnation, cross, and resurrection Christ heals us as members of our evolved species. The "flesh" which the Word became was not some idealized human nature, abstracted from the evolutionary process, but real historical humanity. [88]

By embracing his full humanity, his "earthy" creatureliness, Christ not only has redeemed creation but has also demonstrated to us how we are to live within and otherwise interact with this world. As Thomas Torrance has written: *The incarnation made it clear that the physical world, far from being alien or foreign to God, was affirmed by God as real even for himself. The submission of the incarnate Son of God to its creaturely limits, conditions, and objectivities, carried with it an obligation to respect the empirical world in an hitherto undreamed-of measure.* [89][89]
Thomas F. Torrance, Divine and Contingent Order (Edinburgh, Scotland: T & T Clark, Ltd., 1998), 33.

As a redeemed people we must always remember that we are living in and are a part of a redeemed world. Such knowledge will necessarily transform how we think about God's creation. For a redeemed creation cannot be exploited by utilitarian use, but must be cared for and healed.

From the Editor....

As Rebecca Flietstra stated, the doctrine of the Incarnation offers a resounding confirmation of God's concern for this physical world. In Christian thought, Jesus was the visible sign of an invisible grace. As such, Jesus was a sacrament. Sacramental theology is an important part of Christian thought. In the west, however, this idea has been seriously limited by the fixing of sacraments to a certain number of particular acts. In our journey toward discovering the sanctity of nature, sacramental theology becomes an important element. Although this theology has been limited in the west, it has taken one much more depth and application in the Eastern Church. This contribution of Eastern Christianity to our exploration of the sanctity of nature is offered in our next essay, written by Robin Gibbons.

Dr. Robin Gibbons is the Administrative Director of the Centre for the Study of Religion in Public Life Kellogg College, Oxford University, and the Alexander Schmemann Professor of Eastern Christianity at the Graduate Theological Foundation. He was professed as a Benedictine Monk at St Michael's Abbey Farnborough in England in 1973 and ordained priest in 1979. It was there that he became introduced to the Eastern Church, especially the Byzantine Tradition. He is also an iconographer and one of his major works can be found in the Monastery of Christ in The Desert (Abiquiu, New Mexico).

Chapter 9
For The Life of the World:
A Contribution from Eastern Christianity
Rev'd Dr Robin Gibbons
Centre for the Study of Religion in Public Life,
Kellogg College, Oxford

Defining the boundaries.
The publication of a book in September 2010, <u>The Grand Design</u>, by the Scientist Stephen Hawking occasioned a rash of articles in the British Newspaper, The Times of London.[90] Front headlines on Friday September 3 2010 stated, '*Hawking: Archbishop leads religious response. Faiths take stand against eminent scientist over the role of God*'.[91] This type of headline is all too familiar to British readers, the false dichotomy that society perceives between religion and science is caricatured in this type of statement, and it is a caricature. However two particular articles in the same paper addressed the issue of science and religion to show that there need not, nor in fact be, an unbridgeable chasm between faith and science. In the leader column the Editor wrote:
The ground for religious faith in the modern age cannot be a misguided insistence that science is the path to God: that way lies intellectual chaos. It is more likely to lie in the pull of emotion and- in the title of a famous essay by William James- the will to believe. Because proof of God's existence is ultimately lacking, only a decision of the heart will suffice. As the Apostle Paul defined it, " faith is the substance of things hoped for, the evidence of things unseen". For the believer, that is an irreducible minimum and a stance of sufficient humility to acknowledge and celebrate the power of scientific explanation.[92]

In our explorations of the environment, planet earth, nature, the cosmos, and the issue of sacredness, a divine realm, this insight can help to bridge gaps between these two powerful areas of human concern. The problem lies in a number of areas not least in the different types of language used to describe reality. The dictum of Paul expressed in the editorial reminds all of us that science and religion look at different, but not antithetical things. Far from Stephen Hawking's insistence that there is no place for God in physics and Richard Dawkins equally assertive statement; *'Darwinism kicked God out of biology but physics remained more uncertain. Hawking is administering the coup de grace'*, the search for God seems more alive than ever.[93] One careful insight can illuminate the tentative front line in the supposed battle between these two areas of thought, the Chief Rabbi of Great Britain, Lord Sacks, writing in another article in the same edition of The Times, pointed out that religion needs to cultivate an attitude of admiration and thankfulness towards scientific research, but he also pointed out the necessity of humility in all our academic and intellectual pursuits, for the believer there is more to wisdom than science. *'There is a difference between science and religion. Science is about explanation. Religion is about interpretation. Science takes things apart to see how they work. Religion puts things together to see what they mean'*.[94] This phrase articulates clearly what I believe we can discover in areas of theological endeavor, a different hermeneutic, that of questioning the who, the why and the how of our human existence and that of the mystery of the Divine life with us. This is at the heart of what we mean by the sacred in a Christian viewpoint, not an external transcendent reality but one actively engaged in and with the whole of life.

The Eastern Christian tradition is well placed to help us discover this world of interpretation. Largely untouched by medieval scholasticism, the orthodox and oriental tradition of Christianity has a highly symbolic and metaphorical approach to theological discourse. It is no accident that it is from this tradition that the great spiritual world of monasticism received its Christian format, in this way of perception, the life of the monk becomes the long search for God, a desire to find the word that gives life, a realization of the new creation that has come into being through the incarnation and resurrection of Christ. All that is a distinctively eastern Christian approach to the sacredness of the world, a discovery of it linked together through the incarnation and through the divine immanence makes every created thing a part of what is. For the faithful believer, the questions of science are important and necessary for understanding life, but the questions of religion are different, they ask about the relationship of

God with life in all its aspects. Part of my own search to rediscover the richness of this visual and symbolic language in eastern Christianity starts in the praxis of faith, the living out of faith in celebration, specifically in Liturgy, rightly understood as the 'work' of the people of God and a means of bringing that dimension of the 'other' the holy' right into the visible context of life.

1. Liturgy as a transformative place

One of the first insights we get from the east is the role and importance of the Holy Spirit, the sanctifier, the one who enables all life to become part of the sacredness of God. The worship of the eastern Christian tradition is full of allusions and invocations (epiclesis) to the Spirit. One of the most beautiful prayers is also one of the most used in the worship of the Church and sums up the gift and work of that person of the Triune God; it is well worth quoting in full: *"O Heavenly King, Comforter, Spirit of Truth, Who art everywhere present and fillest all things, Treasury of good things and Giver of life: Come and dwell in us, and cleanse us of all impurity, and save, our souls, O Good One."*

The implicit theology of this prayer is of the penetrating presence of God through the Spirit in all things and all places, both externally and internally. Far from positing a particular fundamentalist religious stamp on the matter of creation, the prayer indicates a concern for relationship rather than origin, an expression of sacredness to be discovered and honored in everything, so that creation is present to us. The question we need to ask is what do we do when we encounter our environment, life forms, and humans? Scientific knowledge leads to discoveries, enlargement of knowledge and therefore of wonder, whilst religious belief engenders a respect for the encounter and a fundamental appreciation of the relationship and dependency between all things and God. This is also at the heart of worship.

In the Divine Liturgy of Saint John Chrysostom, which is the Eucharistic rite used by all Byzantine Churches, whether Orthodox or Catholic , we find these words used during the anaphora (that is the Eucharistic Prayer or liturgy of sacrifice), words said as we ask the Holy Spirit to transform the gifts of bread and wine into the spiritual presence and nourishing gift of Christ: '*when He had come and fulfilled all that was appointed him to do for our sake, on the night on which He was delivered up, or rather delivered Himself up for the life of the world...'* [95] The He is of course Jesus Christ and the implications of this phrase ground the activity, mission, ministry and sacrifice of Christ not only in relationship with those who follow Him but in another and transformative way, with the life of the whole world itself. Even if one has little experience or concern for Christianity, the implication of those particular words, inserted in the central portion of the primary celebration of Christian liturgy, indicates an expression of clear and deep concern for creation, nature and all life and a theological identification with the death and resurrection of Christ. The paschal mystery, which in Christian terms is the redemptive and transformative act of Jesus, brings into focus the Trinitarian understanding of the one God as dynamic and relational action in love, that is the Father and Son's love for each other expressed in the power and activity of the ever present person of the Holy Spirit. The theology of the Holy Spirit places the unseen God right into the heart of nature, continually active and vibrant in creation, bringing to life that which is struggling to give birth, helping make grow that which is living and healing that which is sick and weak .Saint Basil the Great wrote in one of his sermons: ' *for there is not even one single gift which reaches creation without the Holy Ghost (Spirit).'*[96]   His insight places God, not humanity at the center, and in the same sermon he builds on that insight: ' *The whole universe cannot give us a right idea of the greatness of God; and it is only by signs, weak and slight in themselves, often by the help of the smallest insects and of the least plants, that we raise ourselves to Him'.*[97] This insight into the sacrality of all things, even the most insignificant, emphasizes the gift of the Spirit present in all things and all places, and is particularly important within the great tradition of Eastern Christianity with its deeply held pneumatic theology[98] , but also its spiritual teaching about the indwelling presence of God and the human persons capacity to become part of that indwelling light of the transfiguration of Jesus on Mount Tabor, a light of deep compassionate love.[99] Looking into the tradition of this portion of Christianity and this stress on a transformative and transfiguring grace, we discover this is at the heart of the allusion to the light of Tabor, Jesus on the mountain seen as he really is in his glorified and finally resurrected bodily self (see Mk 9.2-8; Mt 17.1-8; Lk 9.28-36; 2 P 1.16-18) . Hopefully a return to 'bodiliness' in this way will develop insights that become important for the future of our concerns and hopes for the world especially with our particular concerns for the environment and all creation.

If, as the Byzantines chant in the Easter hymn, '*Christ is risen from the dead, trampling down Death by death and to those in the tomb he has given life*'[100] , then Jesus Christ who gave his life for the world, places an implicit command on those who follow him in discipleship to give their lives, their all, for the life of the world. If we extend the image of death not only as that of our own mortality, but by extension to the myriad deaths we cause 'nature' to suffer, then the resurrection of Christ empowers those who believe in Jesus Christ to act and restore life to all in the grave, whatever kind of grave that might be. There is a deep and implicit connection between this faith in Christ and creation, another hymn, sung at the Easter service and composed by Saint John of Damascus points this out: '*For it is right that the heavens should rejoice, and that the earth should be glad, and that the whole world, both visible and invisible, should keep the feast*'[101] and again, ' *Now are all things filled with light; heaven, and earth, and the places under the earth. All Creation doth celebrate the resurrection of Christ, on who it was founded*'. [102]

The Christian cannot ever be excused from a deep and fundamental connection with our planet, its resources and all life, especially those aspects of life, nature or living creature of land air or water, which is dependent on humanity for its nurture and survival. It is particularly in worship that the eastern Christian discovers a profound understanding of creation and is given the hermeneutic that can work with scientific knowledge to reshape and alter not only perceptions of our world, but the world itself.

This vision of a new sense of holiness in creation through the resurrection of Christ, introduces a particular topography of sacredness that is no longer fixed to distinctive places, rather is fluid and all encompassing, as well as immediate and personal. Part of this fluidity will be seen in the interpretive tool of a theology of the Holy Spirit, which we will examine more in the baptismal rituals, but as Jerome Neyrey argues in an exposition of worship in John's gospel[103] , there is a definite transition for the Johannine disciples, both in terms of movement away from fixed to fluid space and an understanding of the term my father's home (house) (Jn 14.2) not understood as Temple or location but now understood in terms of relationships with the Divine and each other: '*The model of fluid sacred space urges us to examine how both the person of Jesus and the persons of the group become the sacred space*' [104] . it is also something implicit in a NT understanding of a person and community becoming 'living temple', a personal rather than topographical space, which also means that the locus of Divine activity is everywhere and always, particularly in encounter, and that with all creation not just the human one .Unfortunately as a faith community the Church has not always been in the forefront of this fluid, topographical, vision of sacredness or as we can put it , open concern for reverence towards and understanding of the holiness of our world. Nevertheless there is a deeply rooted theology of not only 'good stewardship', but of 'implicit sacredness' at the heart of creation and in its myriads of aspects which we can particularly discern in the traditions of East and which needs to be brought to the forefront of any debate and dialogue with those who care deeply for the future of life on this planet. It is also a theology that is not at odds with scientific research, its deeply spiritual and symbolic points of reference help inform and elucidate the mystery of God and existence, human animal, avian, aquatic, plant, nature, the universe itself. In the East here is no problem with the search for understanding and explanation, there is no difficulty with evolution, theologically we are in the New Aeon, the new age of the resurrection and its gathering in of all that was and is made into the Kingdom.

2. Renewed Creation: The gift of Christ

Where can one begin to retrieve a deep theology of sacredness within the tradition of Eastern Christianity? There are many places, but in one sense it makes sense to begin an exploration of the liturgical rituals of Initiation, that is of Christian Baptism and the concomitant sacraments (mysteries) of Chrismation and Communion. These are after all the primary rituals that make a Christian; entrance rites of the community and in their ritual activity use basic symbols such as body, light, darkness, oil and water. Not only that but there is a theology at work which stresses the rejection of a disordered (sinful) world created by humanity and the acceptance of a new creation (or perhaps a recreation), one which is already part of the Kingdom of God, not yet visible or accomplished fully in us or our world, but nevertheless present in and to us through these sacred realities. Totally implicit in this theology of a 'new' creation, of the New Aeon, that is the inauguration of the kingdom and reign of God, is the risen presence of Christ through the power of the Holy Spirit, and of humanities potential for transformation and redemption through Christ in the fluid and personalized approach to sacredness, but also of the indwelling sacrality of the 'innocent ones' of creation. These are those who do not need forgiveness or rebirth to mediate the presence and holiness of the Divine world, meaning in some sense everything that is outside of the creation we call man and woman! Eastern Christian theology continually brings us back to that point; it is only the human who has this innate capacity for destruction and sin, placing themselves above the Divine One as the center of life, the rest of God's creation does not.

This is not new teaching, from the 4th century, in the form of a fragment of a longer homily , Dionysius tells us that at the end of creation God blessed (that is made holy) all creation: 'After this the Lord looked upon the earth and filled it with His blessings. With all manner of living things he hath covered the face thereof' but then he goes on the point out that human beings need to recognize the holiness of creation and perceive in these things the action, gift and grace of the Spirit at work: '*And of these Men are not hopelessly blinded, let them survey the vast wealth and variety of living creatures, land animals, and winged creatures, and aquatic; and let them understand that the declaration made by the Lord on the occasion of His judgment of all things is true*' And '*all things, in accordance with His command, appeared good.*' [105] To place this in the context of the baptismal rituals is to acknowledge that in the praxis of theology, the eastern baptismal rites state visually and textually that everything is good, everything mediates the presence of the sacred except sinful man or woman, who being restored to life by these elements in an enactment of death and resurrection, then has the capacity for becoming the very icon of Christ (the true icon of God) whose earthly body is a living temple of the Holy Spirit. Part of the opening and exorcism rite shows this ; '*Put off from him(her) the old man and renew him(her) unto life everlasting; and fill him(her) with the power of the Holy Spirit, in the unity of thy Christ; that he (she) may no longer be a child of the body but of the Kingdom'[106].* This insistence on the sinfulness of human actions is contrasted with the paean of praise to not only the Kingdom but of all creation: *'Great are You , O Lord and wonderful are Your works, there is no word which is sufficient to tell of Your wonders....All the Powers endowed with intelligence tremble before You. The sun sings to You. The Moon glorifies You. The Stars meet together before Your presence. The light obeys You. The deep trembles before You'[107]* .

This unequivocal acceptance of the sacredness of earth and its constituent parts is further strengthened in the following section of the rite which is the long blessing of the baptismal water. This states quite unequivocally that the creature of God 'water' is not evil but, can only be used for harm by those who are evil. In itself it is a number of positive things: *'But do You O Lord and Master of all, show this water to be the water of redemption, the water of sanctification, the purification of flesh and spirit, the loosing of bonds...Wherefore O Lord, manifest Yourself in this water, and grant that he (she) who is baptized in it may be transformed'[108].* It is quite clear that in this ancient rite (as in all eastern baptismal rites) no blame, no negativity is placed on the power, action and fact of water, instead it is seen as a mediating icon of the presence of the Divine One, a manifestation of sacredness. If we stretch this insight to link this with a consideration of the sacredness or otherwise of the natural world, it seems obvious that here is a deep rooted understanding, within the Judaeo-Christian tradition, of the implicit holiness of all matter.

The fact that water is such a powerful symbol of new life can surprise nobody, but that it becomes a means of creating that fluid and relational sense of holiness is important to the wider understanding of fluid, relational sacredness. The connection between the value of the worthiness of creation and value of the human condition is inseparable in this liturgical theology. Basil points this out in one of his Homilies:

'But whence do I perceive the goodness of the Ocean, as it appeared in the eyes of the Creator? If the Ocean is good and worthy of praise before God, how much more beautiful is the assembly of a Church like this, where the voices of men, of children, and of women, arise in

our prayers to God mingling and re-sounding like the waves which beat upon the shore'. [109]

Underscoring this particular theological theme is the next section in the baptismal ritual which is the blessing of olive oil, linking what I describe as a theme of presence and relationship with yet another natural product capable of the deepest symbolic encounter.

Within the hermeneutic of the Liturgical rite and the believing communities participative activity in those forms of worship, texts like these offer an insight into a tradition badly in need of attention in order to effect a full recovery of a deep and integrated understanding of holiness rather than a dualistic religious viewpoint where matter is seen as evil and spiritual as good. It is in relationship with us and God that these elements of world take on a role of mediation and transcendence. After all, in the theology of transfiguration, Christ is transfigured on the mountain, in the world not outside of it! Here is part of the blessing of olive oil which comes after a section of the ritual text illustrating its use and place in the Hebrew story of redemption and holiness: *'Bless this holy oil with the power, and operation and indwelling of Your Holy Spirit, that it may be an anointing of incorruptibility, the armour of righteousness, to the renewing of soul and body...to the deliverance of evil of those who shall be anointed in faith'* [110] It is no accident that oil in this biblical and ritualized sense, has been and still is used for a multiplicity of activities from consecrating the King, anointing the priests but also capable in more bodily terms of being light, food and healing for others. The barrier between the power of the numinous (to use Otto's words) and humanity does not exist in the case of water and oil, as with so many other 'natural elements'. Why? Because they have the capacity to act as a symbol and real conduit of graced holiness between God and human beings. Again we find Basil pointing this out; *'What shall I say? What shall I leave unsaid? In the rich treasures of creation it is difficult to select what is most precious, the loss of what is omitted is too severe...But not a single thing has been created without reason, not a single thing is useless'*[111]. There is a strong sense of proportionality in this eastern view of things, references to creation not only permeates their liturgical celebrations but is continually being referred to in catechetical homilies and letters.

Nevertheless at the heart of it all, and this is connected to the inferences found in the baptismal office, the bridge is that personalized sacrality found in the reconnectivity of sinful-but now redeemed humanity with the risen Christ ,through the power of the Holy Spirit who is the source of sanctification. Gregory Nazianzen puts it succinctly; *'He was baptized as man-but He remitted sins as God-not because he needed purificatory rites Himself, but that He might sanctify the element of water.'*[112]

The liturgical life of the Eastern Church is central to its spirituality, life and mission. In examining the richness of this deep spirituality, which is also understood as a locus of primary theology, *'The Church as koinonia tôn hagiôn-the community of the holy (communion sanctorum) that is built up by the Eucharistic gifts (the sancta or holy things) -and, inseparable from this first aspect, the Church as koinonia of the Spirit, these are central focuses of liturgical interpretation'.*[113] There is a huge corpus of liturgical text, symbols (and praxis) gestures, which makes explicit and deepens this concept of interlinking realities, of heaven and earth. In so many hymns phrases such as, *'today is creation illuminated, today do all things both heavenly and earthly rejoice. Angels and men are intermingled'*,[114] point to the personalized holiness of the community on earth and the work of God through Christ and the Holy Spirit amongst us. Nowhere is the more explicit than in the Eucharistic offering, the title of this essay, *'for the life of the world'* has already linked part of the very deep pattern of sacred realization in Christ to a defined connection, God becomes one of us, heaven and earth become the same locus of divine life although the fullness of this Kingdom is not yet fully achieved! Further on in this great prayer of sacrifice and thanksgiving the following phrase occurs, which the priest sings as the Deacon lifts up the elements of consecrated bread and wine: *'Thine own, of thine own, we offer unto thee, on behalf of all, and for the sake of all'.* In essence this sums up the intermingling we have examined, Christ is both the giver through God the Father of the gifts of material bread, work of human hands, fruit of the earth- and wine, fruit of the vine, but they become transformed through the work of the Holy Spirit, given back to us as our spiritual food and drink. We offer what is given from the earth (thine own) which through the epiclesis of the Spirit is blessed and filled with divine life so that for the believer it becomes the presence of Christ in the offering (of thine own) which brings us into communion, given back so we may partake of them (in all and for the sake of all) and share in Christ (God's) life, that is we become at one with the Divine. Fr Alexander Schmemann brought out much of this in his writings: '*To claim that we are God's creation is to affirm that God's voice is constantly speaking within us and saying to us,*' And *God saw everything He had made, and behold it was very good" (Gen 1.31)...It is redeemed through the incarnation, the cross, the resurrection and ascension of Christ, and through the gift of the Spirit at Pentecost. Such is the intuition we receive from God with gratitude and joy'.*[115] This insight only serves to point out the contribution theology can make to the discourse with science and the context of our needs in examining how the 'sacred' can be a valuable concept and dare one suggest reality. This theological agenda is the part of the key to our problems which trouble our world.

Creation as the icon of God

The theology of the Icon is rich and impressive, in this ancient art form, totally made from natural materials, with nothing synthetic from wooden board to earth pigment paint, a rich and deeply symbolic connection has constantly been made between what is ours here and what in a sense is part of the unseen reality of God. The Icon has had a long history particularly within the Byzantine tradition and is often understood as a means of connection, a vehicle to focus on the sacred. It emerged as a rich art form from ancient times but became accepted as a valid way to image the unseen amongst us. John of Damascus' famous phrase that we can give honor to what has been assumed by the Divine God through the incarnation of Christ, allowed the tradition to create a vibrant art, freed from the taint of idolatry. The icon still plays an enormous part in the liturgy but also in the devotional and personal life of the eastern Christian-and interestingly has been assimilated into the West, although perhaps not quite used in the same way. It carries with it a sense of the sacred but also something which is of grave concern for us. ' *Every icon was a mystery in itself, not only as an image of the divine world but as its real presence on earth*'.[116]		One area that provides a rich source of visual teaching is the Byzantine theology of the Icon; here we have a whole visual book about creation written in paint. The whole process of writing an icon and the developed theology of image yet again links nature, the divine and human relationships together. This of course has its roots in the great debate over the nature of Jesus Christ and the significant attestation that in Him all creation is somehow redeemed and lifted up into the Divine presence. Or in the more kenotic theology of Saint Paul, where a reverse image points out that in Christ, the divine empties out and takes on earthly life. This theological discourse, the interplay between God, in particular the work of the Triune God, is not merely an historical abstraction, the Eastern Church lives out this Trinitarian interplay in its spiritual and pastoral life. In this of course it is not alone, the Christian Church as a whole is getting to grips with the pressing need for dialogue and discourse with a fragmented world in peril, and Catholicism has a bond with eastern theology (particularly through its Eastern Churches) and the hope and expectation of the parousia engendered by the resurrection and ascension of Christ is common to all. But we need to be reminded of the New Aeon, the real 'New Age'.

This brings us back to the theology of icon, often referred to as 'doors of perception'; they are for the Orthodox an understanding of transfiguration, life caught up in reflecting God. Jesus is the true 'icon' of God, where nature and divine life are enmeshed and irreducible, his incarnation brings us into contact with the scared because we become 'icons' of Christ and thus link to God. Fr Alexander Schmemann, one of the greatest of 20th century Orthodox theologians expresses this well; *'In Him(Christ) man, and through man the whole of "nature" , find their true life and become a new creation, a new being, the Body of Christ'*. This is precisely what the theology of the icon hints at, it is a theology of passage and transformation where what is 'old' becomes ever 'new'. This links up with the sacredness of presence implicit through eastern perception, resulting from its strong understanding of the Spirit, again as Schmemann wrote, this is because: *'mankind and creation were called from the beginning to be the Temple of the Holy Spirit and the receptacle of Divine Life'.* [117]

But in all of this what of the living creatures, unlike humanity, different icons of the divine, sacred world, what place can they have in the vision of the sacred, some of whom human-kind has actively demonized and destroyed and others progressively manipulated for their own enjoyment or need without much thought to any consequence? Part of this retrieved insight can be found in the monastic tradition and its counter cultural shifts, but also in the ancient desert tradition of returning to a natural state where the animals become friends and where the true search for God produces the gift of humility, deep compassion for all things accompanied with the outpouring of the gift of tears, for once this develops in a person's life, a whole different view of world is brought about. That is a spiritual path involving individual ascesis, the monastic way (monos seeking God alone) is not for everyone although its witness and example can be. But there are also collective ways of reaching an understanding of the little ones of creation, for is true that in the tradition of daily prayer the living world is being continually referred to in the psalms of Israel, utilized by Christians in their own way, and great reference is made to creatures great and small, valleys, hills, seas, rivers all of which are understood to praise the Lord. And if one takes quite seriously Jesus' own teaching about cleanliness laws, that they are subsumed under a greater Law, that of love of God, neighbor and oneself, and when one links into that neighbor as a wider connection with living things and into God, then they are a true source of Divine encounter. The old proscriptions about categories of animals that are unclean or forbidden or whatever no longer applies in this Christian context, this means that it is incumbent upon us to take seriously our concern for the huge and wicked exploitation of land animals and water creatures. Religion has not got a good track record here. Many Religions have their demonized animals (one can think of the poor old cat in post reformation Protestant Christianity or the snake, serpent, seen as good in some traditions but associated with evil in others), but then they also exalt others and invest them with great significance, like the dove as a sign of peace of God's activity, what we need to do is be all embracing in this.

Unfortunately the Judeo-Christian tradition has not been helpful in history to the animal world and we struggle with our scriptures to find concrete statements to build up a better theology of care, but once again a deeper examination of the Eastern tradition brings out another story, a hermeneutic of care for and understanding of the importance and intrinsic place of animals (and creation) in the pattern and plan of life itself, not as secondary elements under humanity, but as neighbors on the planet. Even if Basil could say that quadrupeds are devoid of reason and is a bit unsure about fish, squids and crabs, he still sees them as exemplars for us, but nevertheless[118], his sermons continually praise the goodness of creation, and he notes, in fairness to his wider perception of God's love for them that; '*At the same time how many affectations of the soul each one of them expresses by the voice of nature. They express by cries their joys and sadness, recognition of what is familiar to them, the need of food, regret at being separated from their companions, and numberless emotions*'.[119] There is a remarkable degree of insight here, in an age when animals were very much less understood. But there it is also a powerful acknowledgment of a more holistic approach to life itself in that far from being 'dumb beasts' they have a complexity about them. Basil sees that they have a far greater capacity for 'subtleness'[120] than humans, dogs for instance can teach us gratitude and loyalty[121] , and all have the capacity for some form of realtionality with humans and God. Certainly Basil and the Cappadocians understood that God understood them, which means they have the potential for sacredness in that conductive sense: '*All bear the marks of the wisdom of the Creator, and show they have come to life with the means of assuring their preservation*'[122].

It is by sharing the insights and teaching of our traditions that a new and greater knowledge can be made available to the future. Religion is not an enemy of science and science is not the end of religious belief: '*the Christian solution requires a change of heart. It seeks wisdom or 'sophia', the third person of the Trinity, as a spiritual guide towards a loving and respectful approach to the world. Sophia seeks the good of many and the reduction of harm to all, both humans and the whole planet*'[123]. The Ecumenical Patriarch Bartholomew I, known for his immense contribution to environmental issues as the 'green Patriarch' calls for the need to engender a 'eucharistic' and ascetic' spirit. He is picking up two themes that his predecessor urged when he asked Christians to observe September 1st as a day of prayer for the environment. His interpretation for 'Eucharistic' implies visioning the world as a 'gift' for which the proper response is gratitude and thanksgiving (Eucharistic) asceticism on the other hand implies sacrifice. Here are his words: '*...but askesis* *signifies much more than this. It means that in relation to the environment, we are to display what the Philokalia and other spiritual texts of the Orthodox Church call enkrateia, self-restraint. That is to say, we are to practice a voluntary self-limitation in our consumption of food and natural resources...Only through such self-denial, through our willingness sometimes to forgo and to say 'no' or 'enough' will we rediscover our true human place in this universe... This need for an ascetic spirit can be summed up in a single key word; sacrifice. This exactly is the missing dimension in our environmental ethos and ecological action'.[124]*

How can we achieve this in the face of so much greed? The model of development in this hermeneutic is similar to that of living in a family, parenting, learning to listen and watch with the eyes and ears of the inner self, to learn when to share and when to conserve, when to say yes but also no! Religion and science need to understand each other's viewpoint in order to help 'locate 'an acceptable solution to problems that at first seems intractable; this is part of the dialogue and work the task of 'parenting' the sacred within our fragile world.

137

BIBLIOGRAPHY

Bartholomew I (Ecumenical Patriarch).Encountering the Mystery: Understanding Orthodox Christianity Today. Doubleday

Cartwright Austin,R 1980. Hope for the Land:Nature in the Bible. Creekside Press

Gibbons,R.2006. House of God; House of the People of God. SPCK London

Gillet, L.1991. Orthodox Spirituality.St Vladimir's Press

Hann,C.2010. Eastern Christians in Anthropological Perspective. University of California Press

Hart,J. 2004. What are they saying about Environmental Theology? Paulist press

Ken Parry (ed). 2010. The Blackwell Companion to Eastern Christianity. Wiley Blackwell

Losky,V. 1991. Mystical Theology of The Eastern Church. James Clarke

MacCulloch,D.2010.AHistory of Christianity.Penguin

Schaff, P. and Henry Ware.2004. Basil the Great, The Hexameron.Homily VI on Creation. In Nicene and Post-Nicene fathers, Volume 8. Basil, Letters and Select works. Second Series. Hendrikson MA

Elizabeth Theotikoff &Mary Cunningham (eds) .2008.The Cambridge Companion to Orthodox Christian Theology. Cambridge University press

From the Editor....

As Robin Gibbons so poetical portrayed, Eastern Christianity has given us an important gift: the notion that the world is sacramental. To speak of the world in this way leads us not only to a deeper sense of the sanctity of nature, but also to a new paradigm concerning humanity's role within nature. At worst, existing views see nature as the temporary resource store for human use and consumption. In an attempt to redeem this view, we have opted to speak of humans as the stewards of nature. This model comes with good intentions, but it has some problems as well. The problems of the stewardship model are both theological and physical. The stewardship model does not fit well into the theological doctrines of creation and providence, and it does not square with the physical realities revealed to us through the scientific discoveries of evolution. To replace this model with a sacramental model offers powerfully new insights into the relationship between God and nature, and equally powerful new models for our human place within nature. Our next essay will address these issues, as John Kerr enlightens us to *This Neglected Faith*.

Dr. John Kerr is a graduate of the Universities of Toronto, Leeds, and Nottingham. He was Athlone Research Fellow at Leeds University and was a Visiting Research Fellow at both Merton College Oxford and the University of California Berkeley's Center for Theology and the Natural Sciences. He taught chemistry, physics, logic, and theology at Winchester College and has been a visiting lecturer at several British and American universities. He was ordained in the Oxford Diocese in the Anglican Church 30 years ago and was one of the founders and second warden of the Society of Ordained Scientists. John was the first Bruton-Rockefeller Resident Scholar, and he now is Episcopal Chaplain to the faculty, staff, and students at The College of William and Mary in Williamsburg, VA.

Chapter 10
This Neglected Faith – World as Sacrament:
a Model for the Sanctity of Nature.

by John Kerr

Introduction

" *All Christian life is sacramental. Not alone in our highest act of Communion are we partaking of heavenly powers through earthly signs and vehicles. This neglected faith may be revived through increased sympathy with the earth derived from fuller knowledge, through the fearless love of all things.*" [125]

FJA Hort, Cambridge scientist and theologian, was one of the most distinguished Anglican thinkers of the nineteenth century. It is in the Appendix to his Hulsean Lectures of 1871, that one finds the remarkable paragraph above.

The three grounds for reviving Hort's "neglected faith" continue to be neglected, even in quite exalted Anglican scholarly circles. For example, in the nearly eight hundred page compendium, "Love's Redeeming Work: the Anglican Quest for Holiness,"[126] co-edited by the Archbishop of Canterbury, there is no reference to the concept of world as sacrament, nor that the study of nature could be conducive to holiness, though such ideas have been at least implicit in a large part of Anglican thinking since the scientific revolution of the seventeenth century.[127]

Moreover, within the prescribed daily Morning Prayer of the Anglican church are canticles, hymns of praise. In one of these, "Benedicite, omnia opera Domini," the whole created order offers up a litany of praise to the Creator: this is a moving expression of "world-as-sacrament." The last creatures in this canticle called to raise their voices in blessing the Lord are humans, and last among them are the priests of religion. Anglican public worship surely is, at least in principle to Anglicans, considered to be a means of inspiring holiness. The cosmic paean of sacramental praise in the Benedicite is only one of many similar themes to be found in the Psalms of the

Hebrew Bible, also part of such worship. It is distinctly odd that the quest for holiness should neglect such a deeply-rooted and distinctive approach to the sanctity of nature.

If there is no spiritual significance, no sacred or numinous property, in the radically-interconnected stuff of the universe, then analysis of its sanctity can have little basis other than the aesthetic. The role of the human as contemplative, of savouring the world's God-given beauty, could suggest that sanctity still could be found, at least in the eye of the beholder. However, (as this essay will stress), such eyes, such beholders, are themselves integrally part of nature.

In other recent Christian writings on environmental ethics by Michael Northcott[128], R.J. Berry[129], and Christopher Southgate [130], and more generally in the works of the late Arthur Peacocke[131], there is a revival of the sacramental model of the symbolic and instrumental sanctity of nature. In the Eastern Christian tradition, such an emphasis has never been lost.

The contemporary youth-oriented "Fresh Expressions" movement of the Anglican Church, much encouraged by Archbishop Rowan Williams, articulates its intense concern for the care of the earth vividly in sacramental language and practice. The integrated offering of praise and service to God and the world, though Christ and in the Holy Spirit, of "ourselves, our souls and bodies in and with the elements in the Eucharist of which they are a part," marks an eloquent recovery of this world view. [132]

The essence of the world-as-sacrament model is that nature, or creation, is acknowledged to possess intrinsic sanctity. Everything that exists praises its Creator in its own right simply by having been brought into and sustained in its unique being, fulfilling thereby the purpose of the Creator. The role of the human, "priest of Creation," is, through deeper knowledge, to discover the particular excellence of all that is and to offer up praise and blessing to the Creator. Unlike the stewardship model, we humans do not stand apart from God or above "nature." We have our place, connected to God, one another, and all that lives, having emerged comparatively recently in the interconnected web of the biosphere's evolutionary history.

The Orthodox, Catholic, and Anglican sacramental model enfolds human praise as one voice raised within that larger praise of the rest of creation. The widely employed stewardship model places humankind in a managerial role outside and over nature, acting responsibly on behalf of an absentee God, standing under God, between God and nature. It does not allow for God's indwelling presence in humankind as part of creation. Nor can it make sense of the extremely limited extent of human existence. Homo sapiens cannot be held to be responsible for, nor to exercise dominion over, 13.75 billion light-years of universe. Nor most aspects of life on earth today. Nor over almost all of earth's history. The sanctity of nature cannot be dependent on our species having declared it.

The term "stewardship" lay available to hand becoming part of the self-advertisement of energy companies among others. Despite its flimsy theological base, almost all of the churches' environmental concern is also premised on "stewardship." The term is something of a "weasel word" and increasingly felt to be suspect. Here on Earth, it has come to mean somewhat less rapacious exploitation now in order to provide benefits for future industrial users and shareholders: there is nothing of sanctity here. What it might mean elsewhere was revealed by the news media's enthusiastic response to data released from an impact probe sent into a deep crater on the moon. The information sent back showed that there was much more water on the moon than had hitherto been known. At once, this was taken to mean that "we" could have a permanent base on the moon and could mine its valuable minerals to send back to meet our growing needs. W.H. Auden's poem "Moon Landing" (1969) expressed his dismay that someday humans could do to the Moon what strip-mining has already done to the mountains and hollows of West Virginia. How prescient he was.

SACRAMENT/ SACRAMENTAL

The sixteenth century theologian Richard Hooker defined sacrament as "an outward and visible sign of an inward and spiritual grace."[133] This definition is widely quoted but requires expanding. The Lutheran Bishop of Lund, Antje Jackelen, wrote, "Properly understood, sacramentality is the radicalization of the idea that a phenomenon is more than it presents itself to be. It also moves beyond a general acknowledgement of the significance of potentiality. Rather than focusing exclusively on the actual, emergence encourages a view of reality as a blend of the actual and the potential. Sacramentality radicalizes this by declaring the potential to be part of the actual."[134] In Celtic and Mediterranean Christianity, sacraments were seen as dramatizations of all nature's redemption, a wider understanding of "salvation history" than is found in almost all post-Reformation Protestant theological understanding.[135] The classical roots of the concept are admirably set out in Professor Anne Loades' historical survey, "Sacrament" in the Oxford Companion to Christian Thought [136]. For purposes of this essay, sacrament will be commended as a traditional way of articulating the spiritual value inherent in nature without reverting to pagan pantheism or New Age nature mysticism and without the anthropocentric overtones and deficiencies of the model of 'stewardship' by dominion.

It is to Arthur Peacocke that we are indebted for this understanding of what sacrament means in a cosmic sense. *"There is in the long tradition of Christian thought, going back to Jesus' own actions and words, a way of relating the physical and personal worlds which avoids any stark dichotomy between them, seeing them rather as two facets of the same reality. This way of thinking is generally denoted by the word "sacramental." In the Christian liturgy, things in the universe – bread, wine, water, oil sometimes, - are taken as being both symbols of God's self-expression and as instruments of God's action in effecting his purposes. This mode of thinking can be extended more widely to the universe as a whole, which can then be seen as both an instrument of God's self-expression, and thus a mode of his revelation of himself, and also the means whereby he effects his purposes in his own actions as agent."*[137]

Peacocke, a physical biochemist, theologian, and Anglican priest, was philosophically a critical realist both with respect to science and religion. He recognized that the technical theological term "sacrament" was uncongenial to many (and not only to agnostics), but thought it provided the least inadequate model for formulating " the duality necessary in our talk about ourselves and about the evolutionary process, avoiding both idealism and the grosser forms of materialism (in the old sense)."

Peacocke's understanding draws heavily on the writings of Hooker: *"God hath his influence into the very essence of all things, without which influence of Deity supporting them, their utter annihilation could not choose but follow. Of him all things both have received their first being and their continuance to be that which they are. All things are therefore partakers of God, they are his offspring, his influence is in them and the personal wisdom of God is for that very reason said to excel in nimbleness and agility, to pierce into all intellectual, pure and subtle spirits, to go through all, and to reach into everything that is. "[138]*

'Creation' is used throughout this essay as a theological gloss on 'nature,' the autonomous system of entities and relationships which is the proper subject matter of science. The world-as-sacrament model is premised on the two components of the doctrine of creation, both creatio ex nihilo[139] and creatio continua, itself a somewhat neglected component of the doctrine of creation. Through the interplay of chance within a law-like structure, there is a perichoresis of mutual ongoing interactions of forces within space-time, matter and energy, life and consciousness and all the processes governing evolution and cosmogenesis: God is immanent, permeating the universe, sharing in its open processes. And God is transcendent.

The Franciscan liberation theologian Leonardo Boff, connects human experience, the mirandum, the experience shared by scientific research and mysticism, with the sacramental beauty, harmony, and mystery of reality. In this, he echoes Einstein who declared that the experience of mystery, "stands at the cradle of true art and true science: whoever does not know it can no longer wonder, no longer marvel, is as good as dead."[140]

Boff explains: *"[The world] reveals God's self in it and is enriched with it. God is also transparent in the world and through the world, and hence the world in its totality and details constitutes a boundless divine sacrament. God is also transcendent vis-à-vis the world in God's character as absolute mystery beyond any imagination and cosmic grandeur. God is in the world and beyond it, continually creating it. "[141]*

"Science arose out of wonder in an effort to decipher the hidden code of all phenomena. Reverence leads to mysticism and the ethic of responsibility. Science seeks to explain the how of things. Mysticism surrenders to ecstasy over the fact that things exist, and it shows reverence to the One revealed in them and veiled behind them. It seeks to experience that fact and establish communion with it. What

mathematics is for the scientists, the spiritual laboratory is for the mystic." [142]

Just such a sense of child-like wonder permeates the Centuries of the seventeenth-century Anglican priest-poet Thomas Traherne. [143] The breadth of reference of the sacramental model may be seen in the spiritual classic by Jean-Pierre de Caussade (1675 – 1751) translated as "The Sacrament of the Present Moment." Time is experienced as a created gift: we live and move and have our being, and therefore may encounter God, only in an experienced present moment and must ever move into God's future at one second per second. To be aware of our relationship to time as gift is essential to "centering" in contemplative prayer. De Caussade saw time as a sacrament, and therefore an aspect of the sanctity of God's creation, to be offered back to God in thanks and praise.

Within this wider sense of nature as sacrament, humanity's role is as priest, blessing and offering back in praise to the Creator what we know of the whole of physical creation, - an "upward" motion - thereby continuing God's original blessing in creating and sustaining all that is. Humanity also is called to a ministry of understanding, healing and preservation – a "downward and outward" movement. In such roles, humans are not inert mediators interposed between God and creation. The modern Orthodox theologian, Elizabeth Theokritoff, notes that, "Cosmic priesthood is not a discrete activity, something we do in addition to living rightly in the world…. It is an attitude: an awareness that all things exist for God's glory, and a response of thankfulness." [144]

Any reflection on the sanctity of 'nature', and particularly of 'creation' as sacrament, must acknowledge the fact that among our sources, rabbinic Hebrew, like biblical Hebrew, has no word for 'nature' as separate from human beings, nor for creation as a finished state. Nor is the word 'sacrament' to be found, though it can be argued that the principle is widely expressed, particularly in the Psalms.

NEGLECTED?

A trend to a reduced, more mechanical, model of sacrament occurred in the West, as it lost sight of the Incarnate Christ as the primary sacrament of God. Sacramentology, as taught to me and others in the churches, was decoupled from both nature and even liturgy. The scope and understanding of sacrament was reduced to working out how many of these means of grace, sanctification and forgiveness should be recognized by the church and devising a philosophical account of how these sacraments might operate. All this despite the broad maxim, unrestricted in its reference, in Catholic sacramentology that the sacraments, "cause by signifying" the operation of God's grace: significando causant.

As Kadavil notes, the whole emphasis rested on 'valid' administration, effecting grace in the recipient. The rest of the ritual was reduced to 'ceremoniae' – none of which generated the magical ex opera operandum. All the rest of the Eucharistic service might make worshipers feel more pious but the 'real' bit was when the canonically ordained priest said the right words in the right order and placed his – it was always 'his' – hands in the right places and at the appropriate moment of consecration. "Sacraments contain and convey grace objectively when administered in the appropriate way, by the appropriate minister, with the appropriate intention." [145]
The thrust of post-Tridentine Catholic sacramentology was an emphasis on getting the rubrics right and carrying out procedures precisely. The recipient remained passive in sacramental celebrations. It was the precise sacramental sign that alone got theological attention. Starting with some general theory of the sacraments, one then "applied" it to marriage, last rites, baptism etc.[146] The great Orthodox scholar Alexander Schmemann observed critically: "In their treatment of the sacrament, it proceeded not from the concrete liturgical tradition and worship, but from its own a priori and abstract categories." In some Protestant denominations, the sacraments were all but utterly neglected as they sundered themselves from Catholicism, in the belief that "All roads lead to heaven that lead away from Rome." The wider implications of sacrament, implicit in the Christian doctrine of the Incarnation were thus neglected because the focus was often, though not always, too narrow, at least in the West.

INCREASED SYMPATHY WITH THE EARTH
        Western Christianity from Augustine and Anselm, and among both Catholic and Protestant theologians from the Reformation to the

present, has largely focused on an anthropocentric, ethical model of redemption. The common assumption was that, in Christianity, salvation is concerned with humanity's personal and social existence. Even then, the expected answer to the question, "Are you saved?" was solipsistic and taken exclusively to mean, "Is your soul saved?" Such a model neglected the existence and destiny of the rest of the universe and our place in it, a not-inconsiderable oversight.

There are splendid exceptions: Francis of Assisi's unsentimental "Canticle of the Creatures" communicated the inseparable need for reconciliation on a cosmic scale with all other creatures, including wolves, with an impulse to praise the Creator for all the cosmic order, including fire. Francis did not separate the cosmic dimension from ethical reconciliation with any humans, including the Muslim armies of Sultan Melek-el-Kamel.

Eastern Patristic Christianity retained a vision of salvation as at once personal and cosmic, social and universal. Humankind was placed at the heart of the physical world as both Microcosm and Mediator.[147] In the writings of the Orthodox St Ephrem of Syria and Alexander Schmemann, the world-sacrament is intimately bound up in the whole liturgical worship of the church and the creative tension between God's transcendence and immanence in creation. The reawakened interest in the West over the last five decades in the theological and spiritual riches of the Byzantine and Greek Fathers has rekindled awareness of the ancient grounds of the concept of world as sacrament.[148]

Christopher Southgate's recent study, "The Groaning of Creation,"[149] affirms the Eastern Christian insight that, *"Man is not a being isolated from the rest of creation: by his very nature he is bound up with the whole of the universe, and St Paul bears witness that the whole of creation awaits the future glory which will be revealed in the sons of God. (Romans 8: 18 -22) This cosmic awareness, never absent from Eastern spirituality, is given expression in theology as well as in liturgical poetry, in iconography and, perhaps above all, in the ascetical writings of masters of the spiritual life of the Eastern Church."* [150]

Sallie McFague writes, *"While Christian sacramentalism derives from an understanding of the Incarnation ("the Word became flesh"), the sense of the extraordinary character of the ordinary or the sacredness of the mundane is scarcely a (uniquely) Christian insight. In fact it is more prevalent and perhaps more deeply felt in some other religious traditions."*

She goes on to assert that traditional sacramentalism acts or has acted as a *"counterforce to two other perspectives on nature within Christian history, one that divorces it totally from God by secularizing it and one that dominates and exploits it."* [151] [152]

DERIVED FROM FULLER KNOWLEDGE

George Herbert, another Anglican priest-poet of the scientific revolution wrote:

"Man is the world's High Priest: he doth present
The sacrifice for all: while they below
Unto the service mutter an assent,
Such as springs that fall and winds that blow." [153]

Herbert draws on the Psalms he recited in daily Anglican worship when he maintains that Nature's tongue is not dumb (Pss 19: 1- 3; 97:6; 98: 7 -9; 148). Though we can't hear the song of creaturely praise, we can know we stand in the midst of the chorus of being. Jurgen Moltmann has written that humans "set loose the dumb tongue of nature"[154] through our thanksgiving, but perhaps Herbert saw further and heard more. When, in the "creation" psalms, thanks are offered up "for the sun and light, for the heavens and the fertility of the earth," the human as priest is thanking God not merely on his own behalf but also in the name of heaven and earth and all created beings in them. This act is truly doxology (meaning, 'giving right glory'.) [155]

Very many of Herbert's contemporaries, including theologians such as the Cambridge Platonists and scientists such as Robert Boyle and John Ray, shared his outlook. Later poets influenced by the science of their day, Coleridge and Wordsworth notably, drew on this sacramental view of the natural world. But it is in the writings of nineteenth-century theologians such as Fr P. N. Waggett (a scientist and priest-religious of the Society of St John the Evangelist),[156] Francis Paget and especially Aubrey Moore (also a scientist and theologian), specifically in relation to Darwinian evolution, that one consistently finds strong, clear advocacy of a sacramental sanctity of nature implied by the Incarnation. [157]

CONCLUDING CAVEATS AND DOXOLOGY

"Everything that hath breath" praises God and the "heavens declare the glory of God" in their own way by being what they are, created as they are, and without human beings and apart from them."[158] If creation needs priests, writes Rasmussen correcting an all-too-human tendency to overstate our importance, they are the four living creatures around God's throne (Rev 4: 6 – 8), only one of which has a human face, acting as our representative worshipper in heaven, whereas the others represent the animal creation with no need of human help. [159]

The New Testament scholar Charles Cranfield, provides a further lyrical but humbling cautionary note: *"The Jungfrau and the Matterhorn and the planet Venus and all living things too, man alone excepted, do indeed glorify God in their own ways; but since their praise is destined not to be a collection of independent offerings, but part of a magnificent whole, the united praise of the whole creation, they are prevented from being fully what they were intended to be, so long as man's part is missing."[160]*

If anything, as one reads in George Herbert, we may go so far as to think of the rest of creation assisting our worship (Ps 148; The human praise follows the worship of all other creatures from the angels downward).

THE FEARLESS LOVE OF ALL THINGS

Tiger, tiger, burning bright
In the forests of the night,
What immortal hand or eye
Could frame thy fearful symmetry?

In what distant deeps or skies
Burnt the fire of thine eyes?
On what wings dare he aspire?
What the hand dare seize the fire?

And what shoulder and what art

Could twist the sinews of thy heart?
And when thy heart began to beat,
What dread hand and what dread feet?

What the hammer, what the chain?
In what furnace was thy brain?
What the anvil? What dread grasp
Dare its deadly terrors clasp?

When the stars threw down their spears
And water'd heaven with their tears,
Did He smile His work to see?
Did He who made the lamb make thee?

Tiger, tiger, burning bright
In the forests of the night,
What immortal hand or eye
Dare frame thy fearful symmetry?

William Blake's frequently anthologized poem, "The Tyger," frames one final challenge to our consideration of Nature's sanctity; that of over-sentimentalizing a morally-ambiguous world.

Variolus Major is a large and extremely elegant virus. It is very efficiently adapted to its habitat, homo sapiens. Smallpox has been a terrible scourge, a 'natural evil,' for millennia: its effect, when introduced into the New World, even accidentally, was devastating. Until microscopy developed to the point where the aesthetic qualities of Variolus Major could be appreciated and a fuller understanding of how viruses function was grasped, smallpox was known only though its hideous symptoms.

The World Health Organization's programme eliminated the smallpox virus from nature in 1977. The last human case 'in the wild', as it were, was Ali Maolin, a Kenyan. Humans have all but removed this species from the world: this may be consistent with the human-centered model of stewardship so prevalent in churches, but the advocates of Deep Ecology, (Naess et al) might view such a destruction of a species for human well-being as 'speciesism' because it destroys the integrity, stability and beauty of a biotic community. How could such an elimination of a species be consistent with a sacramental view of nature? The virus causing the cattle disease Rindpest has just been eliminated. When Rindpest arrived in Africa, some ninety percent of all cattle died, bringing about starvation and vast social disruption. Now this virus too has been eradicated. The difficulties of offering up praise and thanks to the Creator for such disease-causing viruses lies in the deepest levels of theodicy, but if the effect can be separated from the cause, perhaps there are possibilities.

The model of world as sacrament sees other species as created in their own right to give glory to their creator by being what they are. Unsentimental sanctity is inherent in the world-as-sacrament model allows that even disease vectors exist for the glory of God.

An older example may help: unlike the hermit Father Theokristos of Patmos, who befriended them, St Isaac the Syrian was terrified of venomous snakes. Lossky cites a meditation of Isaac's: *"What is a charitable heart? It is a heart which is burning with charity for the whole of creation, for men, for the beasts, for the demons – for all creatures ..... such a man never ceases to pray also for the animals, for the enemies of Truth, and for those who do him evil..... He will pray even for the reptiles, moved by the infinite pity which reigns in the hearts of those who are becoming united with God."* [161]

Such a charitable heart is becoming integrated with itself, acknowledging and being reconciled with the inner "Wild Things" in its own subconscious being, and those of other humans and with the frightening elements in the natural order.

Wendell Berry provides a troubling and intriguing conclusion: *"To live we must daily break the body and shed the blood of creation. When we do this knowingly, lovingly, skillfully and reverently it is a sacrament. When we do it ignorantly, greedily and destructively, it is a desecration. In such a desecration, we condemn ourselves to spiritual and moral loneliness and others to want."* [162]

APPENDIX: A POSSIBLE ISLAMIC PARALLEL

Islam cannot make use of any notion of God as Incarnate. However, Dr Kenneth Cragg noted some parallels in a contribution to Montefiore (op. cit.). He explored the idea of khalifah:

"[M]an is seen as entrusted with the care of the earth, on behalf of God. He is God's vice-regent or vicegerent, exercising "dominion" over the material order, that he may thus be the "servant" ('abd) of the divine law. Man is thus servant-master and is not rightly thought of as either, except in being both. The raw material of his technology is the arena of his adoration of God. He consecrates what he rules and only does either in doing both.

Linked to this is the idea of ayat, or "signs". Nature is a sphere of hints, intimations, disclosures, of divine mercy. In experiencing the physical order ... one is really at the point of the convergence of a physical realm and a spiritual. There are purely empirical causation links but these are the points of awareness of God's goodness and provision..... So all experience is "arresting", i.e. summoning us to discernment and attentiveness. ... Such apperception evokes gratitude. When the "signs" are greeted, worship is released. Thus, in a way, all our awareness is awareness of the "sacramental." The Qu'ran is constantly calling men to thankfulness – a gratitude responsive to a felt significance." [163]

From the Editor...

To see the world as sacred calls us to bear witness to the fact that God has declared nature to serve in litany of praise to the Creator, as well as to serve as an instrument of God's pervasive grace. This invites us to see the sacramental nature of nature. As humans, we have come onto the scene in this most recent of evolutionary eras, possessing the unique ability to recognize the Creator and to claim all nature as being in relationship to that Creator. We alone seem to possess the ability to contemplate the sanctity of nature, and we alone seem to possess the ability to raise nature up in praise to God. Humans are not the stewards ruling over nature in the absence of the Creator. Humans are the High Priest of nature, lifting nature up to the very present God in acts of praise and consecration. John Kerr's essay gives us much to think about.

Kerr's essay ends with an Appendix that offers a possible Islamic Parallel. His essay ending also invites us to begin thinking about how humans should carry out this priestly role. These two ideas move us from Christian thought to Muslim thought, and from theology to ethics. Our next essay brings these two notions together, as Dr. Basil Mustafa discusses environmental ethics from an Islamic perspective.

Dr. Basil Mustafa is the Nelson Mandela Fellow in Educational Studies and Kellogg College Tutor at the Department for Continuing Education, University of Oxford. He is also Bursar at the Oxford Centre for Islamic Studies. Dr Mustafa has been closely involved with a range of activities concerned with the educational needs of ethnic minority groups and with interfaith dialogue. Dr Mustafa is an active member of the Abrahamic Faith Group convened by the Bishop of Oxford and has organized training workshops for RE teachers in state schools. Recently, Dr. Mustafa was involved in leading a conference requested by the HRH the Prince of Wales entitled *Islam and the Environment.*

Chapter 11
Environmental Ethics:
An Islamic Perspective
by Basil Mustafa, Oxford

Part A

The famous 8th Century Muslim jurist, Al Shafie was asked, "What is the proof for the existence of God?" "The Mulberry tree", he replied "its colour, smell, taste and everything about it seems one and the same to you. But a caterpillar eats it and it comes out as fine silken thread. A bee feeds on it and it comes out as honey. A sheep eats it and it comes out as dung. Gazelles chew on it and it congeals producing the fragrance of musk. Who has made all these different things come from the same type of tree?"

This story is illustrative of the Muslim conception of nature as the manifestation of God's presence. It bears close resemblance to the religious study of nature for the better understanding of God; a classical view of what is known as natural theology where nature is conceived primarily as a symbolic system through which God communicates with man (White, 1967). Islam teaches that God has revealed Himself to man both through His creation and revelation, however, the primary sources of Islamic religious knowledge remain the revelations of the Qur'an and the Prophetic traditions.

Rationalism from an Islamic perspective does not mean that reason is superior to revelation, but the rejection of any ultimate contradiction between the two (Al Faruqi, 1984). Muslims view the divine source of the creation and the revelation as ONE, hence they draw harmony and congruence between the two. They see a profound unity in all the diversity of the creation that surrounds us. Above all, they see the 'origin', the 'purpose', and the 'destination' of the creation as inter-related themes, which form the world view of humans and shape the quality of life on earth.

In Islam this message is termed "Tawhid", an Arabic word which means affirmation that there is only one God, the Creator, who is the origin and He only, deserves our praise and worship. The Muslim scripture defines the primary purpose of human creation as

154

the recognition and worship of the One Supreme, Just, Most Gracious and Most Merciful God. As part of fulfilling this purpose, humans were required to cultivate the earth with responsibility and compassion as well as to enjoin justice and maintain the balance in all their relationships with each other and with the natural world.

Islam, which is derived from the root word 'silm' meaning peace, is a way of life based on the willing and humble submission to the Creator. God's revelations and creation breathe peace and inspire tranquillity into human souls. It provides spiritual nourishment that permeates the human existence and helps to inculcate harmony and compassion among people as well as between people and the natural environment. Human beings, due to their multifaceted nature, have a complex set of physical, emotional and spiritual requirements. The ultimate challenge of their creation is to harmonize these various elements into a smoothly functioning holistic living.

Islam, therefore, concerns itself not merely with 'spiritual' matters but with all aspects of human life, blending the spiritual and the mundane in keeping with the reality of the human creation. In order to achieve synthesis, integration and balance within one's life, the individual must exercise the power of his/her will and lead his life according to the values that God has prescribed for his holistic well-being.

The Qur'an refers to the natural environment as the source of our sustenance as well as being an adornment and beauty for human life, *"Nay - who is it that has created the heavens and the earth, and sends down for you [life-giving] water from the sky? With it We cause gardens of shining beauty and delight to grow"* (27: 60). This implies that there is a close connection both material and spiritual between people and the natural world. Nature not only provides our food and shelter, but it also responds to our spiritual needs through the appeal and inspiration of its variety, beauty, order, complexity and connectedness.

Material delight and bodily pleasures are considered natural and sanctioned by Islam on condition they are lawfully attained, moderately used and compassionately shared. Islamic principles offer guidelines on human behaviour and human-nature interaction. The prohibition of inflicting harm or reciprocating with harm is a principle which has become one of the central universal maxims of Islamic law.

155

The prophetic tradition states *"There should be neither harming nor reciprocating harm"*. This principle has become the foundation of many legal provisions that evolved, such as indemnity against damage, prescribing the elimination of and protection against damage. People are expected to empathise with each other and extend compassion to other living things including their natural habitat. The Prophet taught that offering a grain like seed to a bird or a sip of water to a thirsty animal is a rewarding deed, *"There is a reward for (kindness to) every living thing"*. Islamic textual sources contain numerous references that demonstrate a worldview which recognizes the inherent value of non-human life, both animal and plant, and their inter connectedness with human life. *"There is not an animal (that lives) on the earth, nor a being that flies with its wings, but (forms part of) communities like you"*. 6:38

"And in cattle (too) you have an instructive example. From within their bodies We produce (milk) for you to drink; and there are, in them, numerous other benefits for you; and of their (meat) you eat". 23:21

The Prophet stated *"If any Muslim plants a tree or sows a field and a human, bird or animal eats from it, it shall be reckoned as charity for him"*. He also warned against the destruction of life in vain. *"There is no person who kills a sparrow or greater without purpose, save that God Most Glorious and Majestic will hold him to account for it"*. One should add that these teachings have strong connections with the Franciscan mode of charity and the attitude of St Francis of Assisi towards the natural world, especially his love for nature and for animals.[164]

The Qur'an pictures the life of humans as a free competition among them for doing the better, more gracious and nobler deeds. On this account it called the individual a Khalifa, or trustee who acts with authority and responsibility to explore the universe. The world God created is one that fits this moral vacation of humans. It is one in which they are free and effective, where the realization of truth and falsehood, goodness and evil, is actually possible (Al Faruqi, 1984). The concept of human stewardship of the natural environment entails that the values of justice and accountability are the primary principles affecting the relationship between man and the trust which was bequeathed to him. The Qur'an states that "God *commands justice, the doing of good, and kindness to kith and Kin*" (16:90).

The earth's resources and all its life forms are in that sense the responsibility of man to manage and preserve; to use but not to deplete or destroy. This notion is informed by a sense of purpose for which life itself in all its forms has been created on earth. It is also informed by a sense of bond and partnership that joins human beings with nature. Human beings must view themselves as guardians and not masters of this universe. This concept of trusteeship leads to an ethical framework that defines the Islamic rules of behaviour concerning the natural environment and its sustainability. Personal accountability to God is one of the highest values in Islam. In this context it is instrumental in advancing the level of social and environmental responsibility. There is a chapter in the Quran known as "The Bees". It is replete with references to the natural world and where the verses end, it is common to find expressions such as " ...*in this is a Sign for those who are wise*" or "*Verily, in this is a Sign for those who give thought*".

The principles of environmental ethics in Islam encompass the rules of moderation, balance, and conservation, which are the core values of sustainable living (Al-Damkhi, 2008). Muslims maintain that the earth was made malleable for utility, pleasure and comfort of humans. Earth provided the initial constituents of man's creation. "*From the (earth) did We create you, and unto it shall We return you*" (20: 55). The earth was mentioned no fewer than 350 times in the Qur'an and its resources were clearly designated for the benefit of mankind (Al Qaradaghi, 2008). Since these resources are of limited proportions they should be used with moderation for the benefit of all generations. The Qur'an enjoins humans to "*Eat and drink: But waste not by excess, for Allah loves not the wasters*" (7:31). Moderation of consumption was emphasized in many instances where guidelines were laid down by figures of religious authority. Umar, the second successive ruler after Prophet Mohammad visited the market place and observed a companion buying meat on three consecutive days. He asked him why was he buying meat for three days in a row. The companion told Umar that he and his family desired meat every day. So Umar remarked with exclamation

"Should you buy everything that you desire!?"; an exclamation that requires Muslims to exercise self-control over consumption.

One of the moral values of the duty of fasting is to train Muslims on self-restraint and to enable them to temper their desires and moderate their consumption demands. This may be in stark contrast to the excessive consumption demand for food and commodities in Muslim countries today. The paradox between Islamic teachings demanding a human commitment to no excess and no waste and the pursuit of extravagance and wasteful consumption in Muslim societies today will be addressed in the second part of my paper.

Water is the basis of life and it is a precious resource which must be protected from pollution and used in moderation, whether water is scarce or abundant. It was reported that on one occasion, the Prophet observed one of his companions Sa'ad who was using more water than necessary for his ritual ablution. He instructed Sa'ad not to waste water. Sa'ad explained that water was abundant and wondered if it might seem acceptable to use a little extra for ablution. The Prophet explained that even when a Muslim was using the abundant waters of a running river he should be mindful of using sufficient amounts only and avoid wastage. The aim of such an intervention was partly to promote self-consciousness of the need to preserve the precious natural resources and encourage moderate consumption. Water is not just a precious resource, but it is a vital component in the creation of all living things. The Qur'an states: *"ARE, THEN, they who are bent on denying the truth not aware that the heavens and the earth were [once] one single entity, which We then parted asunder? and [that] We made out of water every living thing?" (21:30)*

It is also regarded as a gift from the Creator indicative of His mercy and care for humanity and the Qur'an described rain water as blessing 'blessed water' from God (Abdel Haleem, 1998).

The Islamic tradition teaches that nature was created in perfect ecological balance and harmony. *"Verily, all things have We created in proportion and measure"* (54:79). *"And the earth We have spread out; set thereon Mountains firm and immovable; and produced there in all kinds of things in due balance"* (15:19).

The beauty of nature and its complexity is a testimony of the glory of its creator. Humans are therefore enjoined to maintain the earth's ecological balance while carrying out a responsible exploitation of its natural resources. The human species is just one of many millions of other species inhabiting the earth and all of them have a dependent relationship with others. This calls for certain obligations to arise between the species in order to preserve biodiversity intended for the planet. The accelerated deterioration of the environment we are currently facing is a result of our ignorance and disregard for the delicate balance inherent in the planet's eco-systems. The Qur'an highlights the importance of this balance *"And the sky has He raised high and has devised (for all things) a balance, so that you might never transgress the balance"* (55:7-8).

It follows that all manners of human production and consumption should take into account the broader natural order and balance with the aim of preserving them for future generations.

Conservationists are now advocating the need for a comprehensive environmental account involving not just conservation policies, but the whole range of values underlying environmental protection. The principal value that we have alluded to earlier in this paper and which highlights most expressions of the conservation ethic is that the natural world has intrinsic and intangible worth besides its utilitarian value. The primary sources of Islamic teaching reinforce the ethics of moderate resource use, allocation and protection with the aim of maintaining the integrity and sustainability of the natural world; its habitats, fisheries and biological diversity. This extends also to the conservation of material and energy sources. The protection of soil and the improvement of agricultural land were encouraged by the Prophet of Islam. In one tradition, he advised the faithful to protect the fertility of agricultural land. The protection of water is also mentioned in a number of traditions. It was made forbidden to contaminate sources of stagnant water such as ponds and to discharge sewage into streams of rivers. There are often competing interests of economic prosperity, environmental integrity and social equity which makes the conservation aim more difficult to realize. Conservation efforts are often hampered by low level of public awareness of the issues and their impact both locally and globally on the quality of life.

In the context of environmental ethics, it is useful to reflect on the concept of 'Justice' in Islam. Being just and equitable is one of the ninety-nine attributes of God. When man was created and assigned the role of steward on earth, it was expected of him to carry out two primary responsibilities; first is to worship the Creator and implement His wish for justice to prevail. This was made clear in the instruction given to Prophet David "*O David! We did indeed make you a vicegerent on earth: so judge you between people in truth and justice*" (38: 26). The second duty is to cultivate the earth for the benefit of human life, the Qur'an says "*He brought you out from the earth and has made you cultivate it*" (11:61).

Needless to say that the two duties ought to be carried out in a complimentary manner where the cultivation of the earth's resources must be governed by consideration of equity and fairness which brings into play most expressions of conservation ethics.

In conclusion, the teachings of Islam emphasise that environmental preservation is an ethical and legal requirement based on the following principles (Izzi Dien, 2000):

- The natural environment is God's creation and as such it has an intrinsic value and sanctity besides its utilitarian benefits. Humans are entrusted to preserve its sanctity as a sign of the Creator.
- Several components of the natural environment exist in harmony and submission to the Creator's natural order. Preserving the balance of the natural order is a moral duty and a legal requirement.
- Islamic environmental ethics are based on the principles of justice '*adle*', equity '*ihsan*', and moderation '*wasatiyya*'. Humans are permitted to use the resources of the natural environment with a duty of care and responsibility.
- God entrusted humans with the duty of stewardship which makes the environment their responsibility to preserve for future generations.

Part B

From its origins, Islam clearly offers a conceptual framework for respecting the harmony and balance of our natural world and for ecological stewardship and environmental protection. The sanctity of nature is intrinsic to the idea that it is God's creation. Human beings were tasked with inhabiting and cultivating the earth with an imperative duty of preserving its environment. Having said that, the articulation of an Islamic environmental ethic in contemporary terms is a new development (Foltz, 2003), which in my view has been prompted by two factors. First, is the global environmental crisis now facing humanity and affecting Muslim developing nations in a profound way. Destruction of the Aral Sea in central Asia, desertification of Sub-Saharan Africa, hasty depletion of oil deposits in the Middle East, deforestation and consequently destruction of biodiversity in South East Asia are some well-publicized resource and environmental problems of the Muslim world (Kula, 2001). The second factor is the rising level of public awareness among Muslims of the requirements of their faith and the intensity of their search for a better spiritual well-being. This has given new impetus to practical programmes aimed at preserving the integrity of the local eco-systems in some parts of the Muslim world. It is obvious that Muslims have neglected Islamic principles pertaining to the treatment of the natural environment with a resulting disconnect between religious knowledge and the formulation of environmental policies. Among the many obstacles to practicing Islamic environmental ethics in the modern world are (Nasr, 2003):

1. The dependency of Muslim countries on Western technology and being at the receiving end, these countries are less prepared than the highly industrialized countries to ameliorate the toxic effect of heavy industries. For example in Egypt the industrial wastewater is considered one of the main sources of water pollution because of its toxic chemicals and organic loading (Myllylä, 1995).
2. The mass migration of people from the countryside to the cities, as a result of modern urbanization and industrialization, is evident in many areas of the Muslim world such as Cairo and Jakarta, where

both cities have a population of more that 10 million people. Consequently, this has given rise to significant environmental and health problems.

3. Governments in many Muslim countries are autocratic and the interest of the ruling regime is their first priority. Officials are often hesitant to pursue goals that do not serve those interests. Any environmental movement based on Islamic principles poses a political risk especially if it challenges government policies and plans. The lack of political and civic freedom is therefore a major obstacle.

4. The lack of awareness among traditional Muslim scholars as well as modern Muslim educators about the impact of global warming, destruction of biodiversity and environmental pollution has contributed to the low level of public concern and engagement with environmental issues in Muslim societies.

Although there are some illustrative examples of how traditional paradigms served to encourage a meaningful response to some present environmental challenges, there is still a lot to be desired in terms of capacity building to achieve environmentally sustainable living on a wider scale. According to Hamed (2003), there is enough evidence to suggest that capacity building for environmentally sustainable development cannot be considered in isolation from economic and social development. He asserts that a certain type of human resources is needed in order for any society to develop the capability of managing its environmental resources in a sustainable manner. Raising environmental awareness in Muslim societies therefore becomes an imperative task. This requires the dissemination of the body of knowledge found in Islamic sources relating to Islam's environmental ethics and the conservation of natural resources. Matters of environmental policy are outside the scope of this paper, however underlining the role of education in raising public awareness of environmental issues is important. Engaging the public and gaining the active participation of local people is necessary for initiating change in perception, attitude, and behaviour towards the environment. The role of Islamic teachings, in Muslim societies in particular, can be instrumental in mobilizing people in support of the consumer conservation ethic expressed by the four R's "Rethink, Reduce, Re-use, and Recycle". Education could play a leading role in winning the hearts and minds of young people to adopt a more environmentally friendly way of life. It could enable people to make informed decisions concerning the environment and take responsibility for it. Educational material can provide people with the necessary knowledge and skills they need to replace current practices that are harmful to nature with a more responsible environmental behaviour that supports the goals of sustainable development. The field of environmental education EE in the West has been gaining momentum over the last 40 years and in many ways has been adapting to provide "an interdisciplinary effort aimed at helping learners gain the knowledge and skills that would allow them to understand the complex environmental issues facing society as well as the ability to deal effectively and responsibly with them".[165]

An Islamic perspective of environmental education would stress the social and religious relevance of its content. Thus making it serve the aim of addressing many of the obstacles that are hindering the realization and implementation of Islamic teachings vis-à-vis the environment. This approach could well draw criticism from those specialists who advocate that EE should teach learners "about" the environment in an objective and value-neutral way, thus allowing them to consider various beliefs, values and alternative actions. Teaching people about the environment should not be the end point, but rather the means to work for improved environmental quality based on a rigorous scientific and culturally sensitive approach. It should aim to effect change in our perceptions, values and way of living by generating a personal environmental ethic based on a positive and compassionate view of nature. Nordström defines "environmental education as a multi-discipline line of teaching and learning that educates individuals to become more knowledgeable about their environment and to develop responsible behaviour and skills in order to work for improved environmental quality"[166]. The education process is also expected to facilitate "empowerment" and develop "active citizenship" so that people can address environmental issues in their own communities.[167]

The cultural environment with its most potent elements of faith, language and social history forms an important part of human life and often interacts with the human perception of their natural environment. Therefore, it was argued that strategies of environmental education need to be tailored to the cultural context in order to be effective. Nordström (2008) suggests that the two educational trends of environmental education and multi-cultural education share common characteristics and underlying values. She argues that environmental and multi-cultural education can be used in a complementary manner in schools to achieve a more holistic teaching and learning which will become an effective means for address local, national and global environmental problems.

Muslim scholars have in the past failed to give sufficient attention to affirm environmental values in the content of their Islamic teachings or to emphasise the relevance of those values to a sustainable way of life; with some exceptions relating to animal welfare and water conservation. This situation is improving in that some contemporary Muslim scholars have been discussing the need for educating Muslims about the sanctity of nature and the values that define Islamic environmental ethics. This education, however, is still short of offering a set of skills that will enhance Muslim participation in tackling environmental problems and issues. Sayyed Hossein Nasr has been mentioned earlier in this paper; among other pioneering scholars is Mawil Izzi Deen who has emphasised the role of Islamic environmental education in initiating change. He argued that environmental conservation requires the involvement of the public and proposed that public education in Muslim countries should draw on Islamic values to supplement the provision of environmental education (Izzi Deen, 1990). Other scholars include Abubakr Ahmed Bagader, Othman Llewellyn and Fazlun Khalid. The latter, more recently, published a Teachers' Guide Book for Islamic Environmental Education (Khalid and Thani, 2008). This short guide book was based on the Misali Ethics Pilot Project in Zanzibar. The Misali Island with its coral reefs has suffered environmental degradation, in recent years, due to illegal fishing techniques and over exploitation of marine resources. Community life in that island revolves around Islam and the Misali Ethics Project was initiated to promote conservation at Misali based on the cultural and social context of its community of 1000 fishermen and their families. One can conclude by saying that caring for the natural environment requires all sections of society, parents, teachers, artists, religious scholars, scientists, journalists as well as policy makers to engage in education relating to sustainable living.

References

Abdel Haleem, M.A.S. (1998) Water in the Qur'an, in Abdel Haleem, H. (Ed.) *Islam and the Environment*, London: Ta-Ha Publishers.

Al-Damkhi, A.M. (2008) Environmental Ethics in Islam: Principles, violations, and future perspectives, *International Journal of Environmental Studies*, 65(1), pp.11-31.

Al Faruqi, I. R. (1984) *Islam*, Illinois: Argus Communications.

Al Faruqi, I. R. and Al Faruqi, L. L. (1986) *The Cultural Atlas of Islam*, New York: Macmillan

Al-Qurrah Daghi, A.M. (2008) Islam's View of the Environment: Principles and Applications, unpublished paper, Doha: University of Qatar (in Arabic).

Foltz, R. C. (2003) *Islam and Ecology* (Eds.), Cambridge, Massachusetts: Harvard University Press.

Hamed, S.A. (2003) Capacity Building for Sustainable Development: The Dilemma of Islamization of Environmental Institutions, in Foltz, R.C.; Denny, F.M.; and Baharuddin, A. (Eds.) *Islam and Ecology*, pp. 403-421, Cambridge, Massachusetts: Harvard University Press.

Hungerford, H. R. (2010) Environmental Education (EE) for the 21st Century: Where Have We Been? Where Are We Now? Where Are We Headed?, *The Journal of Environmental Education*, 4(1), pp.1-6.

Izzi Dien, M. Y. (1990) Islamic Environmental Ethics, Law, and Society, in Engel, J. R. and Engel, J. G. (Eds.) *Ethics of Environment and Development*, London: Bellhaven Press.

Izzi Dien, M. Y. (2000) *The Environmental Dimensions of Islam*, Cambridge, UK: The Lutterworth Press.

Khalid, F. and Thani, A. Kh. (2008) *Teachers Guide Book for Islamic Environmental Education*, Birmingham: IFEES.

Kula, E. (2001) Islam and environmental conservation, *Environmental Conservation*, 28(1), pp.1-9.

Myllylä, S. (1995) Cairo – A Mega-City and Its Water Resources, The 3rd Nordic conference on Middle Eastern Studies: Ethnic Encounter and Culture Change, Joensuu, Finland.

Nasr, S. H. (2003) Islam, the Contemporary Islamic World, and the Environmental Crisis, in Foltz, R.C.; Denny, F.M.; and Baharuddin, A. (Eds.) *Islam and Ecology*, pp. 85-105, Cambridge, Massachusetts: Harvard University Press.

Nordström, H. K. (2008) Environmental Education and Multicultural Education- Too Close to Be Separate?, *International Research in Geographical and Environmental Education*, 17(2), pp.131-145.

Palmer, J. A. (1998) *Environmental Education in the 21st Century*, London: Routledge.

White, L. Jr (1967) The Historic Roots of Our Ecological Crisis, *Science*, 155(3767), pp.1203-1207.

From the Editor...

In my experience, most recently affirmed by our dialogue at the Fetzer Institute, the subject of the sanctity of nature, and of its proper use and care, is one that builds bridges among the diversity of people in the world. In this book, we are being led by writings from Christians, Muslims, Jews, and Atheists. We are hearing perspectives from women and men, and from citizens of the United States, Britain, and Israel. The commonality of nature's draw unites us, and the diversity of our experiences and traditions enrich our individual views of nature and its lure. In this next essay, we learn from the rich tradition and heritage of the Africana people, both from Africa itself and from the large Africana presence in the Americas. Drawing from his own experience as well as historical studies, Hugh Page introduces us to the uniquely Africana constructive of the sanctity of nature.

Dr. Hugh R. Page, Jr. is the Walter Associate Professor of Theology and an Associate Professor of Africana Studies at Notre Dame University. His particular research interests are in early Hebrew poetry; the cultural content of ancient epic; theories of myth; African American biblical interpretation; poetry as medium for theological expression; the use of religious traditions and sacred texts in the construction of individual and corporate identity in the Black community; and the role of mysticism and esoterism in African-American, Afro-Caribbean, and Afro-Canadian spirituality.

Chapter 12
Terra Esoterica: An Africana Constructive Spirituality of Nature

by Hugh Page

BACKGROUND

Given the cultural and religious diversity that obtain among the peoples of Africa and the African Diaspora, and the heterogeneous nature of the Christianity encountered within this complex *milieu,* it is impossible to speak of the existence of a single *Africana* perspective on either ecology or the stewardship of nature. For descriptive and analytical purposes, one can, at best: offer a close reading of independent traditions; examine the ways in which selected authors and texts have engaged these themes; and/or describe Africana Gestalten locally, nationally, or internationally. From a theological vantage point, such steps can provide a useful foundation for the creation of 21st century spiritualities that embrace nature in a manner akin to those values and ideals found among *Africana* peoples living on the continent of Africa or those dispersed throughout the world. This essay reflects one such creative intervention. It deploys the religio-philosophical *mélange* known as *conjure* to reconceptualize Black religious history and delimit parameters for one of many possible ecologically centered Africana theologies. When one considers the horrors of the *Ma'afa,* slavery, and segregation – and the role that biblically grounded theologies of subjugation played in justifying them – it is remarkable that so many enslaved and free Africans in the Americas embraced the Gospel, read themselves into the Christian meta-narrative of salvation, and used Scripture to frame their encounter with the cosmos, humanity, and the natural world.[168] The stories of the women and men who did, and of the institutions they created, are well known.[169] The Black Church is a living testimonial to their sacrifices, ingenuity, and hope. It has been a site of resistance and a location in which theological taxonomies reflective of *Africana* realities have been formulated. Of the theological foci central to the Black Church, Gayraud Wilmore (2004:44) notes that:

... three dominant themes or motifs stand out as foundational from the Jamestown Landing to the present. They may be designated as survival, elevation, and liberation.

Nature, it could be argued, has been – and remains – one of the Black Church's many categories of critical engagement. Insofar as the Christian Bible – in its Catholic, Protestant, and various Orthodox canonical forms – is a cornerstone for theological reflection within its constituent denominations some degree of theological reflection on the terrestrial abode of the human family is unavoidable.[170] However, the degree to which nature has been, and continues to be, understood as a distinct *topos* in Black Theology is open to debate. This is perhaps because of what Floyd-Thomas et al. designate the "Black sacred worldview" (2007: 77) undergirding *Africana* life in the Americas.

There is general agreement among those who study *Africana* religious life about how creation has been viewed by many peoples of African descent and conversations about it proceed from several generally held beliefs. The first is that Africans brought with them to the Americas certain fundamental assumptions about the interconnectedness of elements within and spiritual dimensions of the world. Acknowledging her debt to the research of Mbiti, Dube, Olupona, and others, Barbara Holmes (2004: 48) notes that within an African context one finds:… a cosmology that embraces multiple realities without clear demarcation between everyday life and the spirit realm.

The second is that the *Weltanschauung* out of which these assumptions emerged persisted in varying degrees within a variety religious settings, of which the Black Church is but one of many exemplars. It may also be found in African American folklore and popular culture.[171] Many enslaved Africans found sanctuary for contemplation and worship in wooded areas that came to be known as "hush arbors" (Holmes 2004: 82-84). Knowledge of local topography, and adaptability to harsh outdoor conditions, was often requisite for those seeking to escape from Slave States and navigate their way to freedom in the North or in Canada.[172] Maroons throughout the Americas and the Caribbean utilized swamps and other remote areas to provide natural barriers for their communities. Women and men knowledgeable of the healing and *numinous* powers of botanicals

figured prominently in the struggle to end slavery as well as in the
Black imaginary.[173] The use of African labor in New World
agriculture has been seared into the American psyche. The virtual
warehousing of southern immigrants in urban ghettoes in the
aftermath of the Great Migration of the early 20th century, and the
continued sequestering of Black populations in parts of mid-sized and
large cities plagued by poverty and crumbling infrastructure is also
part of the evolving American saga. A piquant verse from B. B.
King's classic "Why I Sing the Blues" is illustrative:[174]

> *I laid in a ghetto flat, cold and numb*
> *I heard the rats tell the bedbugs to give the roaches some*

It could be argued that within a North American *Africana*
context, nature consists of that network of resources and exchange
into which the human family is embedded. This web – potent with
meaning and palpable in its danger, possibilities, and ambiguities – is
the place where *Africana* peoples and cultures exist. Drawing on one
popular West African view of its essential character, it is a *nexus* of
nyama – the power that infuses the cosmos (Doumbia and Doumbia
2004: 5). Clearly, not all persons of African descent would refer to it
in this way. However, many dimensions of *Africana* life are indicative
of the prevalence of such a conception. Christian sensibilities and/or
reticence to speak of such things publically may account for this view
not being more clearly (see Gundaker 1998: 22) articulated and
visible in African American life today.

Given these factors the question arises as to what techniques
might be employed to excavate data related to these ideas from
Africana sources. A reasonable approach would involve conducting a
global historical survey of African and African Diasporan cultural
domains for evidence of ideas about nature. One such study, focused
on African America literature – a pioneering analysis by Kimberly
Ruffin, is soon to be released (Ruffin 2010). Another approach, the
aim of which would be to gather input at the grassroots level, would
involve having cultural insiders reflect on their own experiences. This
is the approach I have chosen.[175] From an applied point of view
another issue is what to do with these data once they are gathered. On
the one hand, they might simply be archived as are other findings

related to Africana cultures. Another would be to apply them, in some way, toward improving the quality of *Africana* and human life once they have been gathered. The second paradigm is consistent with one of the oldest strata of Black Studies. It is also congruent with the goals of that integrative branch of Theology that deals with applied and pastoral matters. For these reasons, I have chosen to utilize it in this decidedly "messy" essay.[176]

A BRIEF ETHNOGRAPHIC MEMOIR

An appreciation of the outdoors was part of my upbringing. I gathered leaves in the autumn, played in winter snows, enjoyed the return of spring foliage, and reluctantly did my share of yard work in the summer heat. In elementary school, exposure to the so-called natural world was part of the curriculum whether it involved trips to the zoo or excursions to Druid Hill Park. We were required to know about famous African Americans – some of whom were Marylanders – their contributions to the observation of nature, their ability to survive it in their quest for freedom, or their capacity to harness it for the common good. The names Banneker, Tubman, and Carver have been part of my vocabulary for as long as I can remember. I learned firsthand how my hometown's location, proximity to the Chesapeake Bay, and geographical placement impacted its relatively mild climate. I discovered as well that its temperate weather was a veil that masked a covertly repressive social landscape. Most of all, I came to know through my own spiritual awakening that: people, things, and events were connected; stray cats could communicate; rain could whisper; and insight can come while sitting in the branches of a mulberry tree thriving in the back alley parking lot of a local bar. All of these experiences were part of the world into which I was enfolded. It had some elements that could be seen, and others that could not. All were tangible and palpable.

Surprisingly, as I was growing up I don't recall hearing, or being involved in, very many conversations – outside of school – about nature and ecology. In sermons at Sharon Baptist Church, my home before becoming an Anglican as an adult, nature was not a *major* touchstone, though I still recall one homily in which our pastor used the story of the inadvertent cutting in half of an earthworm to make a point about the desire of all life forms to persist. Like many congregations that were part of the city's Black Church community, many of its members came from – or had extended family currently living in – what we euphemistically called "the country." To most of my friends, that meant any of a number of places outside of our urban environs: western Maryland, Virginia, North Carolina, etc. Of course, our view was provincial. In our own way, the designation was an attempt to account for not simply their point of origin, but their patterns of speech, cooking styles, life ways, and lore. Stories I heard about "the country" from three of my grandparents, who were born and raised there, and my mother, who had visited often, cast it as a strange and at times dangerous locale to which our family's history was inextricably linked. It was a place where the fabric of Black life was maintained by an odd confluence of sources: e.g., the Bible, the Black Church, agriculture, fishing, domestic service, the underground economy, and a symbiotic relationship with kin who had migrated beyond the Mason – Dixon Line. I remember well stories of one relative who purchased his freedom from a slave owner and two that were preachers. I heard about independent women who left abusive relationships, traveled North, sent money back home, and helped other family members make their way to the city. I learned of the movement of resources (from money to salt-cured hams; from family records to fresh produce) and people along informal networks of exchange.

Many years ago during a hushed conversation in the living room of my maternal grandparents, I recall learning about the African American tarot card reader and spiritual doctor, both of whom had establishments within a three-block radius of their home. Mysterious figures, they had a substantial clientele. Each appeared to be financially independent. Of course, neither was held up as a model to be emulated. I was fascinated by them nonetheless, probably because they appeared to be so far outside of the Protestant Christian *milieu* familiar to me. Only recently have I come to appreciate how central they were to the *numinous* landscape of North Avenue and Mount Street.

I've subsequently learned that card reading and healing through practitioners and material *medica* outside of the Western allopathic mainstream represent a stratum of the *Africana* cultural tradition forged in the Americas that represents a dynamic fusion of

West African, Native American, East European, Christian, Jewish, and other elements (Long 2001: 74, 116- 117, 248; Anderson 2005: 24; Mitchem 2007: 5). Absent an institutional structure, unified architectonic of beliefs and practices, systematic theology, or clergy, its epistemologies and practices are varied.[177] From these, a loose central premise can be distilled: that knowledge, protection, and wholeness can be obtained through the strategic deployment of flora, fauna, physical objects, careful observation, texts, intentions, and words. Everything in the material world has significance, value, and meaning. Many items are polyvalent. To live fully and well, therefore, requires that one be situated squarely within the material world and learn how to read and deploy its elements, some of which have extraordinary power. It is, in a sense, to recognize that there is more to the waking world than meets the eye; to appreciate that there are hidden dimensions to what one sees, hears, tastes, smells, and touches to which one should be particularly attentive. This phenomenon is known by a variety of names. Hoodoo, mojo, conjure, and rootwork are a few of the more well known. There is a burgeoning literature, both academic and popular, on it (see for example Chireau 2003; Yronwode 2002;Bird 2004; and McQuillar 2003).

Deeper meanings, some of which reflect the aforementioned premise, resonate through aspects of *Africana* life in North America that appear on the surface to be completely mundane: e.g., house cleaning, personal care, clothing, the arrangement of bric-a-brac, finding lost items, and even gardening. For some, the origins of these core ideas are unknown. For others, they are known but rarely discussed. I've come to appreciate, through personal experience, in family settings like my own, they come to the surface at surprising times. I've written elsewhere about my experience with tabletops and mantelpieces (Page 2010). Not long after moving into our new home, I received a gift from my godfather: $20.00 to purchase a new broom. Recently, Gundaker's work on property and home fronts has called to mind the prominence of yard work in my family (1998: 14). Moreover, not long before her death, my maternal grandmother leafed through a popular homeopathic *repertory* I'd given my parents as a gift. They were quite surprised when she indicated longstanding knowledge about many of the remedies described therein.

I recall vividly the pride she and the other members of my family had when I set out on my ordination journey, in a sense honoring that part of our familial religious legacy. What haunts me now is what their response might be to my desire to: gather more of the unspoken lore of my elders; query our everyday rituals; and construct an *Africana* spirituality that blends these elements into an inclusive Christian theological *matrix*. Yet, I think that this is exactly where the quest for a Christian spirituality that honors the particularities and ongoing challenges of Black life in North America must begin.

A CLOSING THOUGHT

In establishing a preliminary framework for a liberating and inclusive spirituality whose approach to nature is consistent with the "Black sacred worldview" (Floyd-Thomas et al 2007: 77), the assumption stated earlier – i.e., that nature itself should be seen as those relationships, resources, and systems of exchange upon which peoples of African descent rely from cradle till death and beyond – must be the starting point. Furthermore, the proposition that *conjure* – a loose system of practices related to protection and healing through the manipulation and deployment of power-infused naturally occurring and manufactured items – should be its starting point is reasonable. It is, as well, a plausible framework for pondering larger issues related to *Africana* stewardship, accountability, identity construction, and social structure in a North American setting. Such issues almost always factor into conversations about nature,[178] whether such take place within a traditional Western intellectual paradigm or not. It also serves as an interesting point of departure for Christians to think about changing urban landscapes and eroding infrastructures where crumbling buildings, weeds, and musings on theodicy coalesce. To fashion a spirituality that takes seriously such things would be to reclaim a fundamentally "conjurational"[179] frame of reference as primary and normative for transformational applied Christian theology in Africana contexts. The poem capture this confluence most poignantly for me.

1708
A plowed trash strewn lot
Is all that remains of 1708
The Westwood Avenue home
Where a family told stories, worked, and dreamed
The bricks are gone – the people as well
Do the wildflowers remember?
The streets remain – but for how long
I drive by slowly – wondering …

REFERENCES

Anderson, Jeffrey E. 2005. *Conjure in African American Society*. Baton Rouge, LA: Louisiana State University Press.

———. 2008. *Hoodoo, Voodoo, and Conjure: A Handbook*, *Greenwood Folklore Handbooks*. Westport, CT: Greenwood Press.

Bird, Stephanie Rose. 2004. *Sticks, Stones, Roots and Bones: Hoodoo, Mojo and Conjuringwith Herbs*. St. Paul, MN: Llewellyn Publications.

Carpenter, Delores, and Nolan Williams, eds. 2001. *African American Heritage Hymnal*. Chicago: GIA Publications.

Chireau, Yvonne P. 2003. *Black Magic: Religion and the African American Conjuring Tradition*. Berkeley, CA: University of California Press.-10-

Denzin, Norman. 1997. *Interpretive Ethnography: Ethnographic Practices for the 21st Century*. Thousand Oaks, CA/London/New Delhi: Sage Publications.

Doumbia, Adama, and Naomi Doumbia. 2004. *The Way of the Elders: West African Spirituality and Tradition*. St. Paul, MN: Llewellyn Publications.

Floyd-Thomas, Stacey, Juan Floyd-Thomas, Carole B. Duncan, Jr. Stephen G. Ray, and Nancy Lynne Westfield, eds. 2007. *Black Church Studies: An Introduction*. Nashville, TN:Abingdon.

Gundaker, Grey. 1998. Introduction: Home Ground. In *Keep Your Head to the Sky:Interpreting African American Home Ground,* edited by G. Gundaker. Charlottesville,VA: University Press of Virginia.

Holmes, Barbara A. 2004. *Joy Unspeakable: Contemplative Practices of the Black Church.*Minneapolis, MN: Fortress Press.

Jones, Stacy Holman. 1998. *Kaleidoscope Notes: Writing Women's Music and Organizational Culture.* Walnut Creek: AltaMira Press.

Long, Carolyn Morrow. 2001. *Spiritual Merchants: Religion, Magic, and Commerce.*Knoxville, TN: University of Tennessee Press.

McQuillar, Tayannah Lee. 2003. *Rootwork: Using the Folk Magick of Black America for Love,Money, and Success.* New York: Fireside.

Mitchell, Henry. 2004. *Black Church Beginnings: The LongHidden Realities of the First Years.* Grand Rapids, MI: William B. Eerdmans Publishing Company.

Mitchem, Stephanie. 2007. *African American Folk Healing.* New York, NY: New York University Press.

Page, Hugh R., Jr. 2003. A Case Study in Eighteenth-Century Afrodiasporan Biblical Hermeneutics and Historiography: The Masonic Charges of Prince Hall. In *Yet With A Steady Beat: Contemporary U.S. Afrocentric Biblical Interpretation,* edited by R. C. Bailey. Atlanta: Society of Biblical Literature.

————. 2010. Early Hebrew Poetry and Ancient Pre-Biblical Sources. In *The Africana Bible:Reading Israel's Scriptures from Africa and the African Diaspora,* edited by J. Hugh R.Page. Minneapolis, MN: Fortress.

Pinn, Anthony B. 1998. *Varieties of African American Religious Experience.* Minneapolis, MN:Fortress.

Ruffin, Kimberly. 2010. *Black on Earth: African American Ecoliterary Traditions.* Athens, GA:University of Georgia Press.

Smith, Theophus. 1994. *Conjuring Culture: Biblical Formations of Black America.* New York:Oxford University Press.

Staley, Jeffrey L. 1995. *Reading With A Passion: Rhetoric, Autobiography, and the American West in the Gospel of John.* New York: Continuum.

Still, William. 2007. *The Underground Railroad: Authentic Narratives and FirstHand Accounts*. Abridgement - edited by I. F. Finseth - of the original 1872 ed. Mineola, NY: Dover Publications.

Wilmore, Gayraud. 2004. *Pragmatic Spirituality: The Christian Faith Through an Africentric Lens*. New York, NY: New York University Press.

Wimbush, Vincent. 2000. Introduction: Reading Darkness, Reading Scriptures. In *African Americans and the Bible: Sacred Texts and Social Textures*, edited by V. L. Wimbush. New York, NY: Continuum.-11-

Yronwode, Catherine. 2002. *Hoodoo Herb and Root Magic: A Materia Magica of AfricanAmerican Conjure*. Forestville, CA: Lucky Mojo Curio Company.

From the Editor...

Throughout the essays of this book, our writers have shared perspectives on the sanctity of nature and the human place within it from points of view that are theological and scientific; that come from personal experience and from ethnic and cultural backgrounds; from peering into the vastness of space and looking deeply into the smallest structures of living organisms. The starting point has been very diverse, but the conclusions have offered a rich unity. Indeed, the combination of these perspectives makes a clear case: Nature is sacred. Humans are every-bit a part of nature. We are, as Nancey Murphy states, eco-physical beings. Yet there is something special and unique about our species. Coming along recently on the evolutionary trail, we possess the ability to contemplate the meaning of things, and to recognize a relationship with the Creator. As such, we have a responsibility within nature. This responsibility has been expressed in the concept of stewardship, but as John Kerr illustrated, there may be a better model. All models and metaphors have their limits, but behind them, is the central quest: to take on a role of caring for, and of properly using, nature.

It is one thing to talk about these issues, but quite another to translate that talk into action. Many surveys have illustrated that although a large number of people express concern for the care of nature, a small percentage actually changes life habits to address that concern. To leave our discussion in the realm of theory does not suffice. How do we translate conviction into action? This is the subject to which we now turn in our next essay. Psychologist Janet Swim, along with colleagues Leland Glenna and Brandn Green, offers a case that congregations, and other religious organizations, can serve as a strong catalysis for achieving environmentally sustainable societies.

Dr. Janet Swim is a professor of psychology at Penn State. In the last three years she has become actively involved in efforts to increase psychologists' involvement in understanding the psychological dimensions of climate change. She has done this most directly through her work with the American Psychological Association as the chair of a recently completed task force report on the interface between psychology and global climate change.

Chapter 13
Religious organizations as a social context for achieving a more environmentally sustainable society

by Janet Swim, Brandn Green, & Leland Glenna

Human activity via population growth, consumption, and resource intense technological advances are outpacing effort to address environment threats caused by these activities. According to the Millennium Ecosystem Assessment (2005) "Over the past 50 years, humans have changed ecosystems more rapidly and extensively than in any comparable period of time in human history, largely to meet rapidly growing demands for food, fresh water, timber, fiber, and fuel. This has resulted in a substantial and largely irreversible loss of the diversity of life on Earth." These activities have harsh implications for the survival of many on the planet. The report indicates that the current extinction rate of species is higher than that found in fossil records and is projected to increase to ten times the current rate. Although impacts of degradation of ecosystems occur worldwide, the poorest are the hardest hit with those in regions such as sub-Saharan Africa, Asia, and regions of Latin America facing the greatest problems. These impacts will spread through communities and across borders as people cope with the human induced environmental problems.

Both individual actions and policies are needed to address environmental problems and most support such actions. Altering household behaviors is an important means of intervention given the large environmental impact of household energy use and consumption patterns (Bin & Dowlatabadi, 2005; Dietz, Gardner, Gilligan, Stern, & Vandenbergh, 2009). About 70% to 80% of Americans believe that it is important to engage in a variety of conservation behaviors (e.g., using public transport or carpooling, raising their thermostats in the summer and lowering them in the winter; Leiserowtiz, Maibach,

& Roser-Renouf, 2010). Policies are important to regulate actions and most support government efforts to address environmental sustainability. In 2010, 82% of U.S. residents supported the Environmental Protection Agency and 73% support the EPAs authority to reduce greenhouse gas emissions from utilities and other major industrial polluters (Hueber, 2010). This support reaches across political party lines.

However, beliefs about pro-environmental behaviors are not uniformly or adequately matched by behaviors either by individuals or by market based or government policies. On an individual level, on average across the 17 different types of conservation behaviors about half of those who believe it is important to engage in these behaviors actually do the behaviors (Leiserwitz, et al., 2010). At the national level, the United States does not have a policy to reduce greenhouse gas emissions and it appears that it will be difficult to pass such measures. The problem can be framed in different ways. At an individual level, it can be framed as a lack of correspondence between attitudes and behaviors (Darnton, 2008), at the group level, it can be framed a commons dilemma (Ostrom, Dietz, Dolsak, Stern, Stonich, and Weber, 2002), at a cultural level, it can be framed as problem of the way that culture mediates our understanding of nature and appropriate solutions to environmental problems (Poortinga, Steg, & Vlek, 2002).

In the present essay we consider the role that religious institutions can play in encouraging pro-environmental behaviors in their own organization and among their members. Religious institutions are important to examine because they are relevant to individual, group, and cultural level influences on pro-environmental behaviors. Religious institutions may be especially important for developing motivations, in the form of moral and normative concerns, for engaging in such behaviors (Steg & Vlek, 2009). We focus on moral motivation via one's values because core values associated with pro-environmental behaviors, particularly values that are self-transcendent, are at least theoretically consistent with many religious teaching (Schwartz & Huismans, 1995). Extrapolating from high rates of attendance of churches in the United States we can surmise that experiences had in religious communities are likely influencing both social and individual level values and that these in turn influence possible personal and political actions. Finally, on a practical level, religious institutions have the potential to provide a community where pertinent information will be disseminated and help develop skills to assist in behavior change. Below we first describe self-transcendent values because of their potential to provide a foundation for pro-environmental beliefs and behaviors and because of the potential to be related to individual religiosity and spirituality. Then we consider the role of religious institutions as a place for developing moral communities that may counter a materialistic oriented society. Lastly, we consider aspects of religious communities that can provide practical means of helping individuals enact pro-environmental behaviors.

Values, pro-environmental beliefs, and religion orientations.

Values are conceptions about what is good and desirable (Schwartz, 1994). They are chronic principles that are used to evaluate and promote preferences, actions, and policies. They are more general and stable than beliefs. Via cross-cultural data, Schwartz and colleagues have identified ten core personal values and that fall along two dimensions. One of the dimensions represents self-enhancement values (power and achievement) versus self-transcendence (universalism and benevolence) values. The other dimension represents openness to change (hedonism, stimulation, and self-direction) vs. conservation (traditionalism, conformity, and security). Values of particular relevance for pro-environmental behaviors are those that correlate with concerns about the impact of environmental problems on the self, others (including people in other locations and future generations), and the biosphere (including animals, birds, plants, marine life). Concern about the impact of environment problems on oneself are related to self-enhancement values whereas concerns about environmental problems on others and the biosphere are related to self-transcendent values (Schultz, 2001; Schultz et al., 2005)

These values and environmental concerns have different effects on individuals' willingness to engage in pro-environmental behavior (Milfont, Duckitt, & Cameron, 2006; (Schultz & Lynette C Zelezny, 1998; Schultz et al., 2005; Stern & Dietz, 1994; Swim & Becker, in press; Thøgersen & Ölander, 2002). Self-enhancement values—preferring, for example, social power and being successful and ambitious—and concern about the impact of environmental problems on the self, tend to be associated with less pro-environmental beliefs and behaviors. Universal self-transcendent values—preferring, for example, a world of beauty and unity with nature and concern about the impact of environmental problems on the biosphere, are associated with more pro-environmental beliefs and behaviors. In between, are benevolent self-transcendent values (being helpful, forgiving, and loyal) and concern about the impact of environmental problems on others (i.e., social-altruistic concerns). These values and concerns are less consistently related to pro-environmental beliefs and behaviors but when they are related, they tend to result in more pro-environmental beliefs and behaviors. The pattern of results for values and environmental concerns suggests that as one expands one's focus beyond the self to other people, pro-environmental beliefs and behaviors increase and when one's focus expands beyond anthroprocentric concerns (the self and other people) to biospheric concerns, pro-environmental beliefs and behaviors are the strongest.

The negative impact of self-focused values and concerns on pro-environmental behavior are important to recognize. A popular method of encouraging environmentally responsible behavior is to target self-serving motivations such as monetary or status motivations. This approach can have practical advantages and such tactics can change behaviors. However, this approach also legitimizes self-serving motivations. These motivations can undermine the ability to make large scale changes and changes that require sacrificing for others (Crompton & Kasser, 2009).

Research on values, indicates a complicated relationship between religious orientations and motivation to engage in pro-environmental beliefs. Although perhaps more frequently associated with the conservation-openness to experience dimension of values, religiosity and spirituality are associated with values along the self-enhancement-self-transcendent dimension suggesting that religiosity and spirituality would be relevant to pro-environmental beliefs and behaviors. Specifically, religiosity and spirituality are negatively associated self-enhancement values (power, achievement, and hedonism) and positively associated with greater benevolence

(Pepper, Jackson, & Uzzell, 2010; Saroglou, Delpierre, & Dernelle, 2004). Negative associations with self-enhancement values suggest that more religious and spiritual individuals would be unlikely to endorse ego-centric environmental concerns and as a result would be more likely to engage in pro-environmental behaviors. Consistent with this argument, spiritual transcendence is associated with gratitude and gratitude is associated with less materialism (Diessner & Lewis, 2007). Lack of support for materialism can be important for countering consumerism and the associated tendency to purchase more than is needed and to use more environmental resources than necessary (Kasser, 2002). Associations with benevolence suggest that religious and spiritual individuals would support social altruistic environmental concerns. The focus on human concerns can be seen in environmental statements that many churches have endorsed (e.g., Earth Care Resource Guide, 2008). Many environmental statements focus on social justice issues, helping people who are harmed by environmental problems.

Yet, religiosity and spirituality may not extend to biospheric environmental concerns. Religiosity is negatively associated with universalism and spirituality has been found to be unrelated to universalism suggesting that religious individuals would be unlikely to endorse bio-spheric environmental concerns. Consistent with this conclusion, literalistic beliefs in the bible have been found to be negatively associated with biospheric pro-environmental beliefs (Schultz, Lynnette Zelezny, & Dalrymple, 2000). Religious individuals may be hesitant to work towards environmental sustainability if, for instance, they perceive this work to be grounded in worshiping nature and not God (Hayhoe & Farley, 2009).

There is some suggestion, however, that variation in the way that people understand religious experiences and beliefs may result in more biospheric concerns via changes in the likelihood of endorsing universalistic values. In countries with greater social-economic development, religiosity is more likely to be positively related to universalism (Saroglou et al., 2004). Further, greater endorsement of emotional aspects of religion (community, meaning and values, personal religious experiences), instead of religiosity, are associated with endorsement of universalistic values (Saroglou & Muñoz-García, 2008).

In sum, religiosity and spirituality are likely motivators of pro-environmental behaviors via their negative associations with self-enhancement values and with ego-centric environmental concerns and via their positive association with benevolence and social-altruistic concerns. Yet, the negative or lack of connection universalistic values and likely similar relation with bio-spheric concerns suggests a reason for restriction in the extent to which religiosity and spirituality are associated with pro-environmental behaviors.

Organizations and values.

Studying religious congregations, rather than studying religious individuals, provides insight into the mechanisms of personal values development within a religiously formed moral community. The ontology of a religious community is to be a place of moral reflection that is formed through interaction with ancient texts, theological concepts, rituals, symbols, creeds and ethical ideals which call people to act through an appeal to a higher, or greater, good. If religious communities, and Christian congregations in particular, are places where individuals are shaped in ways which align them with more self-transcendent value orientations they may be important social locations for encouraging pro-environmental behaviors.

The potential role that religious institutions can have on pro-environmental behaviors can be seen via analogy from the sociology of science. Merton (1973) theorized "science as an institution characterized by a system of strong norms and distinct rewards" (Owen-Smith 2006: 65). In particular, Merton (1973) was sought to explain how university science came to be governed by four norms that were unusual in a capitalist economy: universalism, communism, disinteredness, and organized skepticism. Blume (1974) challenged Merton's functionalist assumptions and argued that universities are embedded in social and economic contexts (Frickel and Moore 2006). However, Merton's insights have remained influential as social scientists continue to examine factors that explain why university science tends to produce more basic and non-proprietary research than private sector research organizations. Furthermore, much research has examined whether university science loses its distinctiveness when universities and industries collaborate on research and when universities seek to produce more proprietary goods (e.g. Glenna et al. 2007; Kleinman and Vallas 2006).

Just as the institution of science, and the university as its organizational home, tends to produce public goods while embedded in a capitalist society that privileges the production of private goods, the culture of religious organizations may also promote values and activities that are generally at odds with those of capitalist society. As a result, religious institutions may share some aspects of the dominant culture but also have the potential to be counter-cultural.

Religious organizations may be seen as a locus for promotion of non-materialistic values. First, the sacred texts and symbols of the many religions reinforce ideas of self-sacrifice and generosity. Second, ritual acts in the context of a repeated social gathering may be important sources of beliefs and values (Collins, 2010). Third, as Charles Lemert argues, religion "is the form of social life in which people together, whatever their differences, understand themselves as doubly finite" (Lemert, 2001: 261). This question of human limits is often posited as an essential aspect of the environmental movement and in the development of pro-environmental behaviors.

Using the concept of moral community, religious beliefs and concepts operate to strengthen participation in the community (Graham & Haidt, 2010). Merging these perspectives with the philosophical insights of Hans Joas and Charles Taylor provides an intriguing possibility for explaining the necessary role participation in moral communities plays in developing pro-environmental behaviors. Furthermore, it provides important insight into the logic of studying religious organizations in particular.

Joas (2000) argues that values are generated in experiences of self-formation and self-transcendence. He arrives at this parsimonious explanation through the work of Charles Taylor, who Joas interprets to be saying that moral development necessarily occurs in the context of intersubjective deliberations. We, according to Taylor, learn and create values through participation in communities. This can happen in any type of repetitiously experienced community. Religious communities, however, may be unique places for moral formation because they are contexts for communication and because of normative diffusion (Djupe & Hunt, 2009). Although individuals may not perceive that their places of worship influence their pro-environmental beliefs (e.g., Pew, 2010), other data indicate that they do via the social communication that occurs among members (Djupe & Hunt, 2009). The findings support a model of religious life that is socially contextual.

Practical implications of religious institutions for pro-environmental behavior.

Religious communities can be important for shaping pro-environmental behaviors not only because of their influence on the formation of basic values, but also because they are communities where information is shared and social networks influence behaviors. Religious organizations' formal and informal means of communication can potentially provide a mechanism for enhancing awareness of and discussions about environmental issues and for making connections between values and environmental beliefs and behaviors (Hitzhusen, 2006). Given a tendency for religious and spiritual individuals to endorse benevolent values, religious institutions can potentially build off of these values, for instance, expanding them to other self-transcendent values encompassed by universalistic values and specifically make connections to environmental justice and biospheric environmental concerns. Although most individuals already believe they should engage in a variety of pro-environmental behaviors (as noted above, Leiserwitz, et al., 2010), addressing these values may be important to increase the salience of these beliefs and prioritization of pro-environmental behaviors.

Beyond environmental education, religious organizations can also provide locations for effective interventions given educational structures and social networks within religious organizations. Effective interventions not only address environmental issues and general motivators but also supply behavioral specific information and motivators and help develop skills to overcome psychological and structural barriers to both deliberate and habitual behaviors (Steg & Vlek, 2009). Religious organizations can provide a location for small groups that can help each other reduce energy use and these groups potentially also influence other members via behavioral diffusion through social networks (Bloodhart, Swim, & Zawadski, , 2010;

186

Staats, Harland, & Wilke, 2004; Zawadski, Swim, & Bloodhart, 2010)
.

Summary

Religious institutions have the potential to be effective at encouraging pro-environmental behaviors because of norms within the community and religious practices and teachings that counter self-enhancing and materialistic values and promote benevolence. Further, religious community structures can provide a mechanism for transmitting information and assistance in behavioral skills. However, there are also likely limitations as well as evidenced by their lack of a tendency to value universalism. Educational efforts may need to specifically address universalism and biospheric concerns, develop links between social-altrustic and biospheric concerns, and disseminate practical and normative behavioral information.

References

Bin, S., & Dowlatabadi, H. (2005). Consumer lifestyle approach to US energy use and the related CO2 emissions. Energy Policy, 33(2), 197-208.

Bloodhart, B., Swim, J.K., & Zawadzki, M.J. (2010). Preparing for Environmental Behaviors: A Proactive Coping Intervention as a Means to Promote Behavior Change. Paper under review.

Blume, Stuart S. 1974. Toward a Political Sociology of Science New York: The Free Press.

Crompton, T., & Kasser, T. (2009). Meeting Environmental Challenges: The Role of Human Identity. Godalming, Surrey: World Wild life Foundation - UK. Retrieved from http://assets.wwf.org.uk/downloads/meeting_environmental_chal lenges_ the_role_of_human_identity.pdf

Darnton, A. (2008). Reference report: An overview of behavioral change models and their uses. Goverment Social Research (GSR). Retrieved from http://www.civilservice.gov.uk/Assets/Behaviour_change_refere nce_report_tcm6-9697.pdfDiessner, R., & Lewis, G. (2007). Further validation of the Gratitude, Resentment, and Appreciation Test (GRAT). The Journal of Social Psychology. Vol 147(4), 147, 445-447.

Dietz, T., Gardner, G., Gilligan, J., Stern, P., & Vandenbergh, M. (2009). Household actions can provide a behavioral wedge to rapidly reduce U.S. carbon emissions. Proceedings of the National Academy of Sciences, 106(44), 18452–18456.

Djupe, P. A., & Hunt, P. K. (2009). Beyond the Lynn White Thesis: Congregational Effects on Environmental Concern. Journal for the Scientific Study of Religion, 48(4), 670-686.

EarthCare Resource Guide (2008). Retrieved from http://www.earthcareonline.org/creation_care_websites.pdf

Frickel, Scott and Kelly Moore. 2006. "Prospects and Challenges for a New Political Sociology of Science." Pp. 3-34 in The New Political Sociology of Science: Institutions, Networks, and Power Scott Frickel and Kelly Moore (Eds.) Madison, WI: The University of Wisconsin Press.

Glenna, Leland L., William B. Lacy, Rick Welsh, and Dina Biscotti. 2007. "University Administrators, Agricultural Biotechnology, and Academic Capitalism: Defining the Public Good to Promote University-Industry Relationships." The Sociological Quarterly. 48(1): 141-164.

Graham, J., & Haidt, J. (2010). Beyond Beliefs: Religions Bind Individuals Into Moral Communities. Pers Soc Psychol Rev, 14(1), 140-150.

Hayhoe, K., & Farley, A. (2009). A Climate for Change: Global Warming Facts for Faith-Based Decisions (1st ed.). FaithWords.

Hitzhusen, G. E. (2006). Religion and Environmental Education: Building on Common Ground. Canadian Journal of Environmental Education, 11(1), 9-25.

Hueber, G. (2010). Survey Results – Americans and attitudes about the EPA. Infogroup/Opinion Research Corporation (ORC), (609) 452-5474. Retrieved October 4, 2010 from http://docs.nrdc.org/globalWarming/files/glo_10091501a.pdf.

Kasser, T. (2002). The high price of materialism. Cambridge, MA: MIT Press.

Kleinman, Daniel Lee and Steven P. Vallas. 2006. "Contadiction in Convergence: Universities and Industry in the Biotechnology Field." Pp. 35-62 in The New Political Sociology of Science: Institutions, Networks, and Power Scott Frickel and Kelly Moore (Eds.) Madison, WI: The University of Wisconsin Press.

Leiserowitz, A., Maibach, E., & Roser-Renouf, C. (2010) Americans' Actions to Conserve Energy, Reduce Waste, and Limit Global Warming: January 2010. Yale University and George Mason University. New Haven, CT: Yale Project on Climate Change. http://environment.yale.edu/uploads/BehaviorJan2010.pdf.

Merton, Robert K. 1973. The Sociology of Science: Theoretical and Empirical Investigations. Chicago: University of Chicago Press.

Milfont, T. L., Duckitt, J., & Cameron, L. D. (2006). A Cross-Cultural Study of Environmental Motive Concerns and their Implications for Proenvironmental Behavior. Environment and Behavior. Vol 38(6), 2006, 745-767.

Millennium Ecosystem Assessment, 2005. Ecosystems and Human Well-being: Synthesis. Island Press, Washington, DC.

Ostrom, E., Dietz, T., Dolsak, N. Stern, P.C., Stonich, S. and Weber, E.U. (2002). The Drama of the Commons. Washington, D.C.: The National Academies Press.

Pepper, M., Jackson, T., & Uzzell, D. (2010). A study of multidimensional religion constructs and values in the United Kingdom. Journal for the Scientific Study of Religion. Vol 49(1), 49, 127-146.

Pew (2010). Few say Religion shapes Immigration, Environment Views. Retrieved Oct. 4, 2010 from http://pewforum.org/Politics-and-Elections/Few-Say-Religion-Shapes-Immigration-Environment-Views.aspx.

Poortinga, W., Steg, L., & Vlek, C. (2002). Environmental Risk Concern and Preferences for Energy-Saving Measures. Environment and Behavior, 34(4), 455 -478.

Saroglou, V., Delpierre, V., & Dernelle, R. (2004). Values and religiosity: a meta-analysis of studies using Schwartz's model. Personality and Individual Differences, 37(4), 721-734.

Saroglou, V., & Muñoz-García, A. (2008). Individual differences in religion and spirituality: An issue of personality traits and/or values. Journal for the Scientific Study of Religion. Vol 47(1), 47, 83-101.

Schultz, P. W. (2001). The structure of environmental concern: Concern for self, other people, and the biosphere. Journal of Environmental Psychology. Vol 21(4), 327-339.

Schultz, P. W., Gouveia, V. V., Cameron, L. D., Tankha, G., Schuck, P., & Franek, M. (2005). Values and their Relationship to Environmental Concern and Conservation Behavior. Journal of cross-cultural psychology, 36(4), 457-475.

Schultz, P. W., & Zelezny, L. C. (1998). Values and proenvironmental behavior: A five-country survey. Journal of Cross-Cultural Psychology. Vol 29(4), 540-558.

Schultz, P. W., Zelezny, L., & Dalrymple, N. J. (2000). A multinational perspective on the relation between Judeo-Christian religious beliefs and attitudes of environmental concern. Environment and Behavior. Vol 32(4), 576-591.

Schwartz, S. H. (1994). Are there universal aspects in the structure and contents of human values? Journal of Social Issues. Vol 50(4), 50(1994), 19-45.

Schwartz, S. H., & Huismans, S. (1995). Value priorities and religiosity in four Western religions. Social Psychology Quarterly. Vol 58(2), 58(1995), 88-107.

Staats, H., Harland, P., & Wilke, H. A. M. (2004). Effecting Durable Change: A Team Approach to Improve Environmental Behavior in the Household. Environment and Behavior. Vol 36(3), 36, 341-367.

Steg, L., & Vlek, C. (2009). Encouraging pro-environmental behaviour: An integrative review and research agenda. Journal of Environmental Psychology, 29(3), 309-317. doi:10.1016/j.jenvp.2008.10.004

Stern, P. C., & Dietz, T. (1994). The Value Basis of Environmental Concern. Journal of Social Issues, 50(3), 65-84.

Swim, J.K., & Becker, J. C. (in press). Country Contexts and Individuals' climate change mitigating behaviors: A comparison of U.S. versus German individuals' efforts to reduce energy use. Journal of Social Issues.

Thøgersen, J., & Ölander, F. (2002). Human values and the emergence of a sustainable consumption pattern: A panel study. Journal of Economic Psychology. Special Issue: Social psychology and economics in environmental research. Vol 23(5), 23, 605-630.

Weber, E. U. (1997). Perception and expectation of climate change: Precondition for economic and technological adaptation. In M. Bazerman, D. Messick, A. Tenbrunsel & K. Wade-Benzoni (Eds.), Psychological and Ethical Perspectives to Environmental and Ethical Issues in Management (pp. 314-341). San Francisco: Jossey-Bass.

Zawadzki, M. J., Swim, J. K., Bloodhart, B., & Lenz-Watson, A. (2010). Spreading the eco-message: An applied social networking program to promote environmental behavior in university residence halls. Paper under review.

From the Editor...

It is time for the last word. Throughout this book, our contributing authors have presented engaging arguments for their ideas. I suspect that given the diversity represented on our panel, we would not find total agreement on every point. This is even truer, I suspect, as this book takes on a readership. It is important for us to remember, as David Thomas pointed out at the beginning of our study, that we all employ rhetoric and language to symbolize reality. Our particular employment of language, i.e. the narrative we embrace, is where fine shades of differences emerge. If we can strip away as much of that narrative as possible, we find ourselves more united around a basic experience. Finding ourselves united at that experience, before we try to put too many defining words on it, is where we began this journey. With the guidance of astronomer Jennifer Wiseman, we peered into the vastness of nature, and were left with a feeling of awe. It is that basic feeling that leads us to our particular expressions of what that awe means. From that, all of our theological and scientific explanations of the sanctity of nature emerge. Yet there, in the center of it all, is the awe of nature: Something equally accessible to everyone.

Our last essay may, at first, seem to be another case for a particular narrative of the sanctity of nature; in this case, a Jewish perspective. Indeed, that is where the essay begins, but I have chosen to place it here at the end because it brings us back to the beginning. It returns us to the basic sense of awe, an equal field for all people.

This concluding essay is offered by Noah Efron. Dr. Efron chairs the Graduate Program in Science, Technology and Society at Bar Ilan University, in Israel. He is also President of the Israeli Society for the History and Philosophy of Science, and a member of the Executive Committee of the International Society for Science and Religion. He has been appointed to serve on the Israeli Ministry of Agriculture's committee to evaluate and regulate genetically modified agriculture and invited to participate in Knesset deliberations on human cloning. Efron has been a member of the Institute for Advanced Study in Princeton, a fellow of the Dibner Institute for History of Science and Technology at the Massachusetts Institute for Technology and a fellow at Harvard University.

Chapter 14
"The Wisdom of Everyman"
The Natural, the Sacred and
the Human in Modern Jewish Thought[180]*

One finds among Jews, in modern times as before, a great
variety of attitudes towards nature. This state of affairs is not unique
to Jews. "Nature" may be "the most complex term in the language,"
as the ecofeminist philosopher Val Plumwood observed. It means
different things to different people at different times and different
places. For Jews, this multiplicity may itself be multiplied, as lack of
consensus (indeed, *scorn* for consensus) about what nature is, and
how it ought to be regarded, is stitched into canonical Jewish texts
themselves and valorized by traditional practices of interpreting these
texts.

* Noah Efron

On a Spring Friday, in the early 1560s, a messenger from
Lublin brought to Moses Isserles (c.1530-1572), the young master of
the famed rabbinic academy of Cracow — a letter so withering that as
he read it, he began to tremble. The letter came from Solomon Luria
(1510-1573), chief rabbi and academy master of Lublin, and arguably
the greatest Talmud scholar alive. Weeks early, Luria had sent
Isserles a complicated question of ritual law. Isserles' reply incensed
Luria, who wrote: "I received your reply... and I saw in it sharp
things, [*so* sharp indeed that] I felt as though a razor pierced my
flesh,... The Torah wears sackcloth and mourns [as] you turn to the
wisdom of the uncircumcised Aristotle at every juncture...."
You displayed for me the wisdom of the uncircumcised
Aristotle about the vapors of the land, and so forth, and I said, "Oy,
that my eyes have seen and that my ears have heard, that most of the
delights [you offered] are the words of the impure, and that [these
words] are uttered by Jewish scholars as a sort of *finery* for the Torah,
Lord save us from this great iniquity.
Isserles pondered the letter over the Sabbath, and replied on
Sunday: "First I will respond to what my master made such a

193

commotion about, that I cited in my first letter a statement of Greek wisdom and the Chief of the philosophers...."

But [those rabbis who opposed studying Greek wisdom] were only afraid of learning the cursed Greek books... of metaphysics... because they were worried that Jews would be drawn to their beliefs and would be seduced by their ...wrongheaded notions, but they did not forbid learning the statements of the scholars and their investigations into the essence of reality and into ...nature, because indeed by virtue of these things the greatness of the blessed Creator is revealed... despite the fact that the scholars of the nations of the world said it. For it was already stated in [the Talmud] that "Everyone who relates a bit of wisdom, even if he be of the nations, is called a wise man." [181]

Isserles was persuaded that the wisdom, or philosophy, of nature is ecumenical. This, he believed, was because nature itself is ecumenical. To learn about nature is to learn about God, Isserles believed, and to do so, one does not have to be a Jew. Nature shares her secrets with the wise of all religions; it is universal and ecumenical.

One of Isserles' students, David Gans, built his career on this insight. Upon completion of his Talmud studies in Cracow, Gans made his way to Prague, and was there when Rudolph II chose it as the capitol of the Holy Roman Empire, watching with delight as the city became home to many of Europe's best natural philosophers, artists and writers.

Three times during the spring of 1600, David Gans made the slow, eighteen-mile trip from his home in Prague's *Judengasse* to Rudolf's splendid summer estate in Benatky. Gans came at the invitation of Europe's most renowned astronomer, Tycho Brahe, and stayed each time for the better part of a week. Brahe had recently established a temporary observatory at the summer palace, and persuaded Johannes Kepler, Johannes Müller, and other talented, young scholars to join him there. Gans used his visits both to observe the heavens and to observe some of Europe's greatest astronomers observing the heavens. The work they carried out, Gans wrote, produced "great things the likes of which in our days we have never seen nor have we heard, and our forefathers did not tell us, and we did not find them written in the books of the Jews or the nations of the world." Gans believed that knowledge of nature and only knowledge of nature could provide an intellectual bridge, a common tongue, common endeavor and common cause, between peoples of different confessions.

This view was endorsed by Gans celebrated contemporary, the Maharal of Prague (the towering scholar who would later be linked in legend to the Golem of Prague). Maharal asked of natural philosophy:

Why do they call it 'Greek wisdom'?, because if it is intended to explain the realities of the world, is it not so that this wisdom is the wisdom of *every man*. [182]

As Maharal and Gans and other Jewish intellectuals of the age saw it, natural philosophy – the study of nature — was different from the other great disciplines – theology, law, metaphysics – because it supplies knowledge that is important, enduring and true for *every* man, Catholic, Protestant, Moslem or Jew. It is this unique trait, Gans thought, that made natural philosophy – science – so terribly important in an age of religious wars. His approach to science can perhaps be called *natural irenism* or *natural ecumenism*: A belief that God – in his beneficence — made nature in such a way as to allow its peaceable study by people of very different beliefs. It was no coincidence, in Gans' view, that Jews, Christians and Moslems had often achieved in the observatory the sort of companionable intellectual collaboration that eluded them everywhere else.

This view of Gans, that one approaches the study if nature as a *human being*, not as a Jew, is part an enduring intellectual tradition, one that at some times had great appeal for Jews, and at others almost none at all. Maharal himself denigrated the study of nature as mere stuff about stuff:

The importance of an [intellectual] attainment varies according to the importance of the subject. And certainly everything depends upon this. If a person labors and becomes wiser than all the ancients [about nature] ...there is no doubt that this is considered nothing compared to a small [intellectual] attainment concerning the hosts of the heavens. [183]

Elsewhere, Maharal was even more explicit:

It is not appropriate to call someone who knows about material things "wise," just as a shoemaker is not called "wise," even though this is [a sort of] wisdom too. Therefore, only the person who studies holy matters [is called wise], and this is called wisdom. [184]

When a Jew studies nature, according to Maharal, he studies the realm of the gentile, and he becomes expert in precisely that which he aspires to transcend. Nature, according to Maharal, is universal because it is profane and profane because it is universal.

Maharal and Gans agreed, then, that nature is a matter of *human* concern more than it is a matter of particularly *Jewish* concern. But they disagreed about what this fact meant. To Maharal, it meant that nature was of little interest, its value merely practical. To Gans, it meant that nature could bring together in common cause people of different faiths, comrades in awe over God's creation. To Maharal, nature was a token of the base animality shared by all humans. To Gans, nature was a corridor to the elevated spirit shared by all humans. To Maharal, nature was mundane. To Gans, it was hallowed.

Together, these two views are opposing aspects of a single tradition that saw as nature's signal trait its universality, its ecumenism, its catholicity. In the four centuries that have passed since the two men conferred in Prague, this tradition has resurfaced in varied guises in many times and places. That this is so comes as no surprise. These four centuries have been tumultuous ones for Jews. Over their course, the relationships between Jews and the rest have been continually defined, refined and then redefined. From the vantage of *Jewish* history, the modern era has been a centuries-long meditation on social ontology – on what was often called "the Jewish Question" – that produced a variety of arrangements for Jewish minorities, many uneasy. In this context, the appeal is obvious of a view of nature that seemed to offer a realm in which Jews, Protestants, Catholics, Moslems, and everyone else were, in end, simply human beings. For this reason, this view resurfaced again and again among Jews, from David Gans to Stephen Gould. Baruch Spinoza, for instance, embraced such a view, observing in his *Tractatus Theologico-Politicus* that:

In examining natural things we strive, before all else, to investigate the things which are most universal and common to the whole of nature – viz., motion and rest, and their laws and rules, which nature always observes and through which it continuously acts.[185]

For Spinoza, it is precisely in this striving for the universal and common that the theological lesson of nature lay. Spinoza believed that the laws of nature ought to guide the framing of our politics and even, in the end, our theology.

And, as I said, this view was enthusiastically embraced by many Jews in the generations after Spinoza. At the end of the 18th century, and beginning of the 19th, many *maskilim*, enthusiasts of what has been called the Jewish Enlightenment, adopted a more positive view of science, much like Gans' (Gans, in fact, became a sort of hero for early *maskilim*). Like Gans, many *Maskilim* took it almost as a religious duty to study the ways of nature, because in so doing the *Jewish* intellectual becomes simply an intellectual. Leaf through the pages of *ha-Me'assef*, the journal of the *Maskilim*, and you will find scientific treatises translated into Hebrew and presented with the sort of care that, until then, was lavished on religious texts alone. *Maskilim* too were drawn to nature for the ecumenism it offered.

But it was in the late 19th and especially in the 20th centuries that Jewish enthusiasm for such a view of nature, and for the *natural ecumenism* it allowed, reached its zenith. It is well known that during these years, Jews turned to the study of nature in remarkable numbers, with passion and capacious success.[186] Especially in the 20th century, everywhere in the West where science thrived, Jews played a vastly oversized role. This was especially true in Germany, the United States and the Soviet Union, but Mainz, Moscow and Manhattan happened, on a smaller scale, in England, France, The Netherlands, Austria, Italy, Poland, Hungary, in short in every Western country with ample measure of both Jews and science. And the inroads Jews made in the sciences in these places, achieving a surprising level of regard, were matched by the inroads science made among Jews, many of whom came to regard the study of nature with respect bordering on awe and devotion bordering on the spiritual.

This is especially noteworthy, because for most of these Jews, awe of nature was the only spiritual devotion they would ever profess. No good prosopography of this remarkable cohort of Jewish scientists has yet been attempted, but the great majority, and nearly all of the most successful among them, were non-religious Jews (what the renowned biographer Isaac Deutscher memorably called "non-Jewish Jews").[187] For at least some of these scientists, the careful study of the laws of nature was their only spiritual exercise.

These modern Jewish scientists find a place in the tradition that runs through Gans and Spinoza and Moses Mendelssohn, the *maskilim*, Samson Raphael Hirsch and many others, scholars who found in nature value and values they found nowhere else. As they saw it, in studying nature, people could avoid the schismatics and factionalism that plagued so many other pursuits. Nature and its study offered rare concord and this concord offered the possibility of real human fellowship. The British mathematician Hyman Levy wrote in his best-selling, 1932 *Universe of Science*, that the purpose of studying nature:
is to reach a system capable of being isolated from the subjective world of every individual member of the human race, ... a set of statements acceptable to all.... The formulations of science... are statements invariant with respect to the individual.188

Six years later, as Nazis across the sea denounced quantum theory as "degenerate, Jewish science," the Jewish sociologist of science, Robert K. Merton wrote (in an affecting essay entitled "Science and the Social Order") that: "It is a basic assumption of modern science that scientific propositions "are invariant with respect to the individual" and groups. ... "One sentiment which is assimilated by the scientist from the very outset of his training pertains to the purity of science. Science must not suffer itself to become the handmaiden of theology or economy or state.

The biologist and writer Benjamin Gruenberg had much the same to say in his *Science and the Public Mind*: Orthodoxies of all sorts, whether religious or political, whether moralistic or intellectualistic, are inimical to the spirit of inquiry [into nature]....[189] An appreciation of the development of science as a great cooperative enterprise of mankind is likely to promote solidarity and to make each individual feel a sense of unity with his fellows.....[190] Science is a means of broadening the sympathies and cultivating tolerance toward other groups, races, nationalities, tastes and philosophies.[191]

Such a view was not unique to Jews, of course, but was famously advocated by the most famously Jewish scientists of the 20th century. Albert Einstein, who proudly declared himself a disciple of Spinoza, held such a view. It was with such a view in mind that Robert Oppenheimer, speaking in 1953, intoned Bishop Sprat's description of the "noble" aims of the Royal Society in studying nature "not to lay the Foundation of an *English, Scotch, Irish, Popish or Protestant* Philosophy; but a Philosophy of *Mankind,*" an aim that Oppenheimer found only eight years after Nazi surrender resonated with "a haunting sense of its timeliness."

And the same view was behind the outrage felt by physicist Richard Feynman when he met Jews who took a particularly *Jewish* interest in how nature works: *One day, two or three...young rabbis came to me and said, "We realize that we can't study to be rabbis in the modern world without knowing something about science, so we'd like to ask you some questions... In the Talmud it say you're not supposed to make fire on a Saturday, so our question is, can we use electrical things on Saturdays?" I was shocked. They weren't interested in science at all! The only way science was influencing their lives was so they might be able to interpret better the Talmud! .. It really was a disappointment. Here they are, slowly coming to life, only to better interpret the Talmud. Imagine! In modern times like this, ...the only way they think that science might be interesting is because their ancient, provincial, medieval problems are being confounded slightly by some new phenomena.*[192]

What galled Feynman was the failure of the Rabbis to see that science is hostile to "provincial, medieval problems," and that nature offers escape from, not an aid to, the sort of parochialism they stood for.

And, of course, such a view was precisely what Stephen Jay Gould championed when he wrote that: Science tries to document the factual character of the natural world, and to develop theories that coordinate and explain these facts. Religion, on the other hand, operates in the equally important, but utterly different, realm of human purposes, meanings and values.[193]

The natural world, Gould believed, was the stuff of stuff, the stuff of fact, the stuff of demonstration, the stuff about which rational, right-thinking people would ultimately agree. The natural world is the stuff that can bring Chinese, Russians and Americans, blacks and whites, Jews, Christians and Moslems into companionable cooperation in a laboratory. Wars may be fought over conflicting human purposes, meanings and values; but coalitions are created through the study of nature. "Science has no borders," as people fond of Gould's view are fond of saying, because nature has no borders. And nature has no religion.

Fifteen years ago, I had the fine fortune of finding myself in a small seminar room at Harvard's science center, as Gould spoke about butterflies and Nabokov (what eventually gelled into an essay called "No Science without Fancy, No Art without Facts: The Lepidoptery of Vladimir Nabokov," that appeared in *I Have Landed*.) Gould had not long before rediscovered Nabokov's specimens in Harvard's Museum of Comparative Zoology. He described them lyrically, with passion and poetry. As he did, his spirit seemed to expand. He described them with love. In this, Gould was of a type that I suspect is familiar to all of us, a fiercely secular scientist who finds sanctity in nature, without having proper words, or proper concepts to describe it. (The handiest example of this type is Charles Darwin, who completed *Origin* with the most religious of irreligious determinations, the most spiritual of spiritless statements: "There is grandeur in this view of life, with its several powers, having been originally breathed into a few forms or into one; and that, whilst this planet has gone cycling on according to the fixed law of gravity, from so simple a beginning endless forms most beautiful and most wonderful have been, and are being, evolved." Gould adored Darwin, and quoted this sentence as a benediction.)

Seemingly, there is a paradox in the views of Gould (and Feynman, Oppenheimer, Einstein, Merton, Levy and the multitudes they represent). They freely speak of nature, and experience nature, in a spiritual idiom, at the same time, they insist that nature's most important characteristic is that it is inimical to spirit. They find great value in nature, while insisting that nature brooks no value. They are reverent about nature because nature demands no reverence. They consecrate themselves to the study of nature, because nature is indifferent to their devotion.

Considered more carefully, the paradox disappears. To Gould and many others leading back to Gans and before, nature is in its way laden in spirit *because* nature is devoid of spirit. The "sanctity" of nature arises from its human implications, from its natural irenism or ecumenism, by the fact that it could know humans only as humans (the laws of nature would not distinguish a Jain from a Jesuit), and could be known by humans only as humans. Gans knew that it was the stars that gave him license to join Tycho Brahe at Emperor Rudolf's summer palace, and that nothing less than creation itself could afford that opportunity. And Gould knew that butterflies under glass could forge a fellowship and bond of spirit between Nabokov, whose parents were wealthy Russian notables, and himself, whose grandparents were coarse Russian Jewish laborers. The line that connects Gans to Gould is peopled by countless other Jews along the way.[194] To them, nature sheds a light by which Jews, like Christians, Moslems and everyone else are people, seeking to make their way together in a world they share. It is no surprise, perhaps, through a recent history that has seen ghettos, exiles, quotas, forced conversions and briefly extermination camps, such a light holds considerable attraction for some Jews. It is in this light that many Jews find grandeur in nature, and, though they may not express it in this way, a rare and precious sort sanctity.

Conclusion

From the Editor...

From my own experience, I can certainly echo Efron's words. As demonstrated in this book project, as well as other programs and organizations in which I have been involved, nature is a true bridge. The subject of nature has brought together the diverse contributors of this book, and allowed us to forge a meaningful community. Exploring the sanctity of nature is important to all of us, because we are all an integral part of nature itself.

Although we are at the end of this study, we find ourselves at another beginning point. Coming to terms with the sanctity of nature, and of our human place within it, is a first step. Translating that discovery into a guiding compass for our treatment and use of nature is the next challenge. It is a challenge that has implications for many things that are a burgeoning part of our lives: conservation; technology; bioethics; fuel production; etc. Even as we conclude this study, it is important for us to continue the dialogue. Once we own the sanctity of nature, then we must own the implications of the sanctity of nature. Our journey is just beginning!

NOTES:
TEACHERS EDITION
About the DVD

The companion DVD included with the "Teachers Edition" provides valuable additions to this book. On the DVD, you will find:

1. Video interview responses from our participants on various questions that were posed to them. The video portions of this DVD were under the direction of Martin Ostrow, director of the film Renewal.

2. Video discussions with each participant individually concerning areas of their own unique areas of expertise.

(Footnotes)

[1] Alexander, Denis (2008) Creation or Evolution: Do We Have to Choose? Oxford: Monarch Books, p.181.

[2] Plato on Rhetoric, http://www.americanrhetoric.com/platoonrhetoric.htm

[3] Murphy, Nancey C. (1994) *Reasoning and Rhetoric in Religion* . Trinity Press International

[4] I probably took this from a Wikipedia article. At other places throughout my essay where I include some specific information like this without attribution, you may consider that it also comes from Wikipedia or some internet source like it. It will not enhance this paper, or my scholarly reputation, to repeatedly insert Wikipedia footnotes. Where I think my internet sources are important to include, I will insert them; otherwise, just consider the source. I have little doubt about the accuracy of my references to information like the Trivium and Quadrivium, or I would not be using it.

[5] Aristotle (1954) The Rhetoric and The Poetics of Aristotle. Introduction by Edward P.J. Corbett. Modern Library College Editions. (paperback). There are several translations of Aristotle available. I cannot read Greek, so all of my references to The Rhetoric and The Poetics of Aristotle are from this or a similar English language version. At the elementary level of my summary here, the translation variances among different versions are immaterial.

[6] Burke Kenneth, A Grammar of Motives (1945), U of Cal Press. Other Burke benchmark studies in rhetoric include A Rhetoric of Motives (1950) U of Cal Press, First Paperback Ed. 1962; The Rhetoric of Religion: Studies in Logology (1961).

[7] Joseph Campbell (1949). *Hero with a Thousand Faces.* Princeton U P.

[8] Burke, *Rhetoric of Religion.*

9 Kenneth Burke (1966), *Language as Symbolic Action: Essays on Life, Literature, and Method*. U of Cal Press.

10 http://tpmmuckraker.talkingpointsmemo.com/texas-textbook-hearings/2010/09/

11 Michael Zimmerman, "Overturning the Texas School School Board Madness? It's Possible," posted online at Huffington Post, Sept. 14, 2010. http://www.huffingtonpost.com/michael-zimmerman/overturning-the-texas-sch_b_715360.html

12 Walter R Fisher (1987). *Human Communication As Narration*. U of S Carolina Press. Paperback.

13 Starting with Sally McFague (1975). *Speaking in Parables*. Fortress Press

14 http://christianethicstoday.com/wp/

15 Fisher, Ch. 3, "Narrative as a Paradigm of Human Communication," 57-84. The five presuppositions of the Rational-World paradigm are all listed on p. 59. Fisher's survey of the "root metaphors" of the many different philosophies of what it means to be human, leading to his argument that the narrative paradigm subsumes all them, is found on p. 62. The five presuppositions of the Narrative Paradigm are listed on pp. 64-65. If you think my capsule summary is complex, try reading the whole chapter. Much of Ch. 3 unpacks these dense ideas, and connects them to Burke's rhetoric.

16 Sallie McFague, *Life Abundant*(Minneapolis: Fortress Press, 2001) p.18

17 Ibid, p.20

18 Oikos is the root of the "eco" in economy and ecology

19 Ursula Goodenough, *The Sacred Depths of Nature* (New York:Oxford University Press, 1998) p. xvi-xvii

20 Andrea Cohen-Kiener.*Claiming Earth as a Common Ground*. (Vermont: Skylight Paths Publishing, 2009) p. 32-33.

21 Ibid, p. 27

22 Ibid, p 36-40

23 Sidney Schwartz, "Exploring Religion, Social Justice and the Common Good", in The Reconstructionist (Fall, 2000), p 26-27

24 Ibid, p 27

25 Ibid, p 28

26 Ibid, p 29

27 From the Song *I Feel Like Travelling On*

28 From the Song *This World is Not My Home*

29 http://www.biologos.org/resources/albert-mohler-why-does-the-universe-look-so-old

30 http://www.time.com/time/world/article/0,8599,443800,00.html

31 http://www.christianitytoday.com/ct/2010/october/3.18.html?start=5

32 Ross, Hugh, 2009. More Than a Theory. Baker Books, p 159

33 Ibid, p 79

34 Behe, Michael, 1996. Darwin's Black Box: The Biochemical Challenge to Evolution. Free press, p 233

35 Isaiah 55:12,13

36 http://biologos.org/blog/on-coming-to-peace-in-the-family-of-god/

37 Psalm 19:1-4

[38] Dobzhansky, Theodosius. 1973. *Nothing in Biology Makes Sense Except in the Light of Evolution*. The American Biology Teacher 35: 125-129.

[39] Vermeij. 2010. The Evolutionary World: How Adaptation Explains Everything from Seashells to Civilization. New York: Thomas Dunn/St. Martin's Press.

[40] Vermeij. 2010. The Evolutionary World: How Adaptation Explains Everything from Seashells to Civilization. New York: Thomas Dunn/St. Martin's Press.

[41] Goldstein, Jeffrey. 1999. Emergence as a Construct: History and Issues. Emergence 1: 49-72.

[42] Kauffman, Stuart A. 2008. Reinventing the Sacred: A New View of Science, Reason, and Religion. New York: Basic Books.

[43] Forrest, Barbara and Paul r. Gross. 2004. Creationism Is Trojan Horse: The Wedge of Intelligent Design. New York: Oxford University Press.

[44] The full text of The Christian Clergy Letter can be accessed at" http://blue.butler.edu/~mzimmerm/Christian_Clergy/ChrClergyLtr.htm.

[45] The Evolution Weekend 2011 web page can be accessed at http://www.evolutionweekend.org.

[46] Bill Moyers Apologizes to James Watt for Apocryphal Quote. Entertainment News, 9 February 2005. Accessed at http://www.entertainment-news.org/bill-moyers-apologizes-to-james-watt-for-apocryphal-quote/ on 27 August 2010.

[47] White Jr., Lynn. 1967. The Historical Roots of Our Environmental Crisis. Science 155: 1203–1207.

[48] . McFague, Sallie. 1993. The Body of God: An Ecological Theology. Minneapolis: Fortress Press.

[49] . Darwin, Charles. 1859. On the Origin of Species by Means of Natural Selection, Or the Preservation of Favoured Races in the Struggle for Life. London: John Murray.

[50] I'll be writing throughout from a Christian perspective, but what I say will be relevant to varying degrees to both Jews and Muslims.

[51] Frans de Waal, Good Natured: The Origins of Right and Wrong in Humans and Other Animals (Cambridge, MA: Harvard University Press, 1996), 13-20.

[52] De Waal, Good Natured, 17.

[53] Arthur O. Lovejoy, The Great Chain of Being (Cambridge, MA: Harvard University Press, 1936).

[54] See Mary Midgley, Beast and Man: The Roots of Human Nature (Ithaca, NY: Cornell University Press, 1978).

[55] This is Daniel C. Dennett's parody in Freedom Evolves (New York: Viking, 2003), 1. But see Harold Bloom's argument to the effect that the "the real American religion is and always has been in fact . . . gnosticism." in The American Religion: The Emergence of the Post-Christian Nation (New York: Simon and Schuster, 1992), 49.

[56] This was fifteen years ago, for my account in Warren S. Brown, Nancey Murphy, and H. Newton Malony, eds., Whatever Happened to the Soul: Scientific and Theological Portraits of Human Nature (Minneapolis: Fortress, 1998). Since then two particularly helpful books have appeared: Raymond Martin and John Barresi, The Rise and Fall of Soul and Self: An Intellectual History of Personal Identity (New York: Columbia University Press, 2006); and Joel B.

Green, Body, Soul, and Human Life: The Nature of Humanity in the Bible (Grand Rapids: Baker Academic, 2008).

57 The arguments here focus largely on Paul's anthropology. The best source, I believe, is James D. G. Dunn, The Theology of Paul the Apostle (Grand Rapids, MI: Eerdmans, 1998), chap. 2, sec. 3.

58 Notice that there is a linguistic shift here from "souls" to "minds." Either term is a fair translation of Descartes's Latin or French. For Thomas the mind was equivalent to the rational soul (intellect and will). For Descartes, everything of which we are conscious, including sensations, is a function of the mind, and all of the other faculties (such as the ability to move) are attributed to the body. Earlier translations of Descartes's writings used "soul," but as this term has increasingly taken on religious connotations, translators have come to prefer the word "mind" in most contexts.

59 H. Wheeler Robinson, The Christian Doctrine of Man (Edinburgh: T. & T. Clark, 1911). While Robinson's account of Old Testament teaching struck a blow against dualism, it did not support physicalism directly since Robinson interpreted theories of human nature in terms of his idealist philosophy.

60 Rudolf Bultmann, Theology of the New Testament, vol. 1 (New York: Scribner, 1951).

61 Oscar Cullmann, Immortality of the Soul or Resurrection of the Dead? (New York, Macmillan, 1958).

62 Dunn, The Theology of Paul the Apostle, 54. Dunn attributes the aspective/partitive account to D. E. H. Whitely, The Theology of St Paul (Oxford: Blackwell, 1964).

63 Alwyn Scott, (2004). "A Brief History of Nonlinear Science." Revista del Nuovo Cimento 27/10-11 (2004): 1-115; quotation at p. 4.

64 15. Francis Heylighen, "Self-Organization of Complex, Intelligent Systems: The ECCO Paradigm for Transdisciplinary Integration." Integral Review, (2011)

65 Austin Farrer, The Freedom of the Will, The Gifford Lectures, 1957 (London: Adam and Charles Black, 1958), 57-60.

66 See, for example, Robert Van Gulick, "Who's in Charge Here? And Who's Doing All the Work?" in John Heil and Alfred Mele, eds., Mental Causation (Oxford: Clarendon, 1995), 233-256.

67 For two approaches, see Nancey Murphy and Warren S. Brown, Did My Neurons Make Me Do It?: Philosophical and Neurobiological Perspectives on Moral Responsibility and Free Will (Oxford: Oxford University Press, 2007); and Alicia Juarrero, Dynamics in Action: Human Behavior as a Complex System (Cambridge, MA: MIT Press, 1999).

68 This is John Mustol's term, developed in his "Physical Humans in an Ecological World: The Implications of a Physicalist Anthropology for Christian Ecological Ethics" (ThM thesis, Fuller Theological Seminary, 2010).

69 Freeman Dyson, Disturbing the Universe (New York: Harper & Row), 250.

70 Mustol's work "converted" me from my despair regarding Christian ecological ethics. Were it not what I learned from him I probably would not have agreed to participate in this project.

[71] Mustol, "Physical Humans," 57; quoting Sallie McFague, A New Climate for Theology: God, the World, and Global Warming (Minneapolis: Fortress, 2008), 92; and Holmes Rolston, III, "Kenosis in Nature," in The Work of Love: Creation as Kenosis, ed. John Polkinghorne (Grand Rapids: Eerdmans, 2001), 51, 52-3.

[72] James Wm. McClendon, Jr., Ethics: Systematic Theology, Volume 1 (Nashville: Abingdon, 1986), 89.

[73] Diogenes Allen, Philosophy for Understanding Theology (Atlanta: John Knox Press, 1985), 39-59.

[74] De Waal, Good Natured, 146.

[75] I can no longer find the reference for this quotation.

[76] This is Lutheran theologian Ted Peters's whimsical term.

[77] Except, of course, for sin.

[78] Cyril of Alexandria, On the Unity of Christ, trans. John Anthony McGuckin (Crestwood, NY: St. Vladimir's Seminary Press, 1995), 109-110.

[79] Larry W. Hurtado, Lord Jesus Christ: Devotion to Jesus in Earliest Christianity (Grand Rapids, MI: William B. Eerdmans Publishing Company, 2003), 2-3.

[80] Athanasius, "Letter LX. To Adelphius, Bishop and Confessor: Against the Arians." http://www.ccel.org/fathers2/NPNF2-04/Npnf2-04-114.htm#P10078_3609272 (accessed January 29, 2004).

[81] Henry M. Morris, "Luke 1:31 conceive in thy womb" in The Defender's Study Bible (Grand Rapids, MI: World Publishing, Inc., 1995), 1084.

[82] Irenaeus, "Against Heresies," Book III, xxii, 1.

[83] Gregory of Nazianzus, "Letter 101.The First Letter to Cledonius the Presbyter" in On God and Christ: The Five Theological Orations and Two Letters to Cledonius, trans. Frederick Williams and Lionel Wickham (Crestwood, NY: St. Vladimir's Seminary Press, 2002), 158.

[84] International Human Genome Sequencing Consortium, "Finishing the Euchromatic Sequence of the Human Genome," Nature 431 (October 21, 2004): 931-945.

[85] For a discussion on proteomics, the difficult task of cataloguing of all human proteins, see Nature Insight: Proteomics in Nature 422 (13 March 2003): 193-235.

[86] George L. Murphy, "Christology, Evolution, and the Cross" in Perspectives on an Evolving Creation, ed. Keith B. Miller (Grand Rapids, MI: Eerdmans, 2003), 385.

[87] Gregory of Nazianzus, "Letter 101.The First Letter to Cledonius the Presbyter" in On God and Christ: The Five Theological Orations and Two Letters to Cledonius, trans. Frederick Williams and Lionel Wickham (Crestwood, NY: St. Vladimir's Seminary Press, 2002), 158.

[88] George L. Murphy, "Christology, Evolution, and the Cross" in Perspectives on an Evolving Creation, ed. Keith B. Miller (Grand Rapids, MI: Eerdmans, 2003), 380.

[90] Hawking, S and L Mlodinow.2010.The Grand Design: New Answers to the Meaning of Life. Bantam Press.

[91] The Times of London. Friday September 3 2010. thetimes.co.uk, No 70043 p1

92 Ibid Editorial p2

93 Ibid Leader p1

94 Ibid article', Even great science tells us nothing about God.p27

95 Text taken from Joseph Raya & José De Vinck, 1995. Byzantine Daily Worship. National Publishing Company Philadelphia. p. 283

96 Schaff, P. and Henry Ware.2004. Basil the Great, The Hexameron.Homily VI on Creation. In Nicene and Post-Nicene fathers, Volume 8. Basil, Letters and Select works. Second Series. Hendrikson MA.p.35 Chapter xxiv

97 Ibid p.89

98 Pneumatic (from pneuma-spirit) refers to a theology of the Holy Spirit

99 This refers to the Transfiguration of Jesus on Mount Tabor, one of the great orthodox feasts of light, a hint and foretaste of our own transfiguration in Christ. (feast August 6th)

100 Hymn for Pascha text taken from, Service Book of the Holy Orthodox-Catholic Apostolic Church, translated by Isabel Hapgood (amended by Fr Robert Gibbons for local use).1996.Antiochian Orthodox Christian Archdiocese of North America. Englewood NJ. p. 225

101 Ibid, Canon for Holy Pascha, St John of Damascus. Canticle I., Tone I,p.227

102 Ibid, Canticle III p. 228

103 Neyrey.J.H. 2007.Give God the Glory. Ancient Prayer and Worship in Cultural Perspective. Eerdmans, Cambridge UK

104 Ibid p.204

105 Roberts, A. and James Donaldson.2004.Ante-Nicene Fathers Volume 6. Works of Dionysius. Extant Fragments. Hendrickson Publishers, PeabodyMA USA. p.91 (200.265.)

106 The Office of Holy Baptism. Service Book.p.275

107 Ibid p.277

108 Ibid p. 278

109 Schaff 2004.Basil Hexameron. Homily IV. Upon the gathering of the waters. P75

110 The Office of Holy Baptism Service Book p279

111 Schaff.2004.Basil.Hexameron.Homily V. The Germination of the earth.p77

112 Schaff.Volume 13.Gregory of Nazianzen.Fourth Theological Oration. P309

113 Schulz, H J.1986. The Byzantine Liturgy. Pueblo. New York. P196

114 From the Great Compline Litiya ,stanza for Byzantine feast of the Epiphany

115 Thomas Fisch Ed. 1990. Liturgy and Tradition: Theological Reflections of Alexander Schmemann. St Vladimir's Seminary Press. New York.pp 98-99

116 Fedotov, G.P.1960.The Russian Religious Mind. Harper Torchbooks, New York. P33

117 Fisch P77

118 Schaff.Basil Homily IV. Upon the gathering of the waters.p95

119 Schaff.Basil.Homily VIII.The creation of fowl and other animals.p 95

120 Ibid p. 104

121 Ibid p. 104

122 Ibid p. 104

123 Dorricot, J.2005.Science and Religion.SCM Study Guide. SCM Press, London. p241

[124] Ecumenical Patriarch Bartholomew I. Archbishop of Constantinople and New Rome. Sacrifice: The Missing Dimension. Address at the closing ceremony of the Fourth international Environmental Symposium. June 10, 2002, Venice, Italy.

[125] Hort, F.J.A., The Way, the Truth and the Life, MacMillan, London, 1908, p.213.

[126] Rowell, G., Stevenson, K., and Williams, R., Love's Redeeming Work: the Anglican Quest for Holiness, Oxford UP, Oxford, 2001.

[127] Runyon, T., The Sacraments, in Wainwright, G., ed. Keeping the Faith: Essays to Mark the Centenary of Lux Mundi, SPCK, London, 1989, pp 209 – 224.

[128] Northcott, M., The Environment and Christian Ethics, Cambridge UP, Cambridge, 1996, pp 131 sqq.

[129] Berry, R.J., ed., Environmental Stewardship, T&T Clark, London, 2006

[130] Southgate, C., The Groaning of Creation, God, Evolution and the Problem of Evil, Westminster, Knoxville, 2008.

[131] Peacocke, A.R., A Sacramental View of Nature, in Man and Nature, Montefiore, H. ed, Collins, London, 1975.

[132] 8Croft, S., Mobsby, I., and Spellers, S., Ancient Faith, Future Mission: Fresh Expressions in the Sacramental Tradition, Seabury, NY, 2010.

[133] Hooker, R. The Laws of Ecclesiastical Polity, London, 1601

[134] Jackelen, A., Creativity through Emergence: a Vision of Nature and God, in Proctor, J.D., ed., Envisioning Nature, Science and Religion, Templeton Press, West Conshocken PA, 2009, p 197-198.

[135] 11White, J.F., The Sacraments in Protestant Faith and Practice, Abingdon Press, Nashville, 1999.

[136] Loades, A., Sacrament, in Oxford Companion to Christian Thought, eds, Hastings, A., et al, OUP, Oxford, 2000. P. 634.

[137] Peacocke, A.R., Evolution: The Disguised Friend of Faith?, Templeton Press, West Conshohocken PA, 2004, p. 48.

[138] Laws of Ecclesiastical Polity, III.ii.8, II.ii.1 and V.xlviii.2.

[139] Burrell, D.B., Cogliati, C., Soskice, J.M. and Stoeger, W.R., Creation and the God of Abraham, Cambridge UP, 2010.

[140] Einstein, A., The World as I See It, in Einstein, Ideas and Opinions, Based on Mein Weltbild, trans. Bargmann, S., ed. Seelig, C., Crown Publishing, NY, 1954 p11.

[141] Boff, L., SSF, Cry of the Earth, Cry of the Poor, trans., Berryman, P., Maryknoll, NY, Orbis, 1995, p. 143, 147-148.

[142] Boff, op. cit.

[143] Margoliouth, H.M., Thomas Traherne: Centuries, Poems, and Thanksgivings, revised ed., OUP, Oxford, 1972.

[144] Theokritoff, E., Creation and Priesthood in Modern Orthodox Thinking, J. Study Religion, Nature and Culture, Ecotheology, 10, 3 December, 2005 pp 344 – 363.

[145] The Medieval Theologians, p178

[146] Kadavil, M., The World as Sacrament: Sacramentality of Creation from the Perspectives of Leonardo Boff, Alexander Schmemann and Saint Ephrem, Peeters, Leuven, 2005.

147 Thunberg, L., Microcosm and Mediator: The Theological Anthropology of Maximus the Confessor, 2nd ed., Open Court, Chicago, 1995.
148 Philokalia on the Prayer of the Heart
149 Southgate, C., op.cit.
150 Lossky, V., The Mystical Theology of the Eastern Church, James Clarke, London, 1957, pp 110 – 111.
151 McFague, S., The Body of God: An Ecological Theology, SCM, London, 1993.
152 Clifford, A., Religion and the Environment,: An Ecological Theology of Creaturely Kinship, J. Religion and Society, Supplement 3 (2008)pp 1 – 11.
153 Paternoster, M., Man: The World's High Priest, Fairacres, Oxford, 1978.
154 Moltmann, J., God in Creation, SCM, London, 1985, p 71.
155 Rassmuson, L., Symbols to Live By, Ch. 14 in Berry, R.J., ed., Environmental Stewardship, T&T Clark, London, 2006 pp 181 – 182.
156 Waggett, P.N., The Scientific Temper in Religion, Longmans, London, 1905, inter alia.
157 Gore, C., ed, Lux Mundi, John Murray, London, 1889: q.v. Ch. 2: Moore, A. The Christian Doctrine of God, and Ch. 10, Paget, F., Sacraments.
158 Moltmann, J., God in Creation, SCM, London, 1985 , p. 71.
159 Rasmussen, L., op. cit.
160 Cranfield, C.,E. B., Some Observations on Romans 8: 19 – 21, pp 224 – 230, in Banks, R., ed., Reconciliation and Hope, Eerdmans, Grand Rapids, 1974, quoted in Rasmussen, p.181, op. cit.
161 Lossky, V., The Mystical Theology of the Eastern Church, James Clarke, London, 1957, pp 110-111.
162 Berry, W., The Gift of Good Land, North Point, San Francisco, 1981, p 281, in Berry, R.J., ed, Environmental Stewardship, p. 187.
163 Montefiore, H., Man and Nature, Collins, London, 1975, pp 162 – 4.
164 Catholic Encyclopaedia at www.newadvent.org
165 Hungerford, 2010, p.2
166 Nordström, 2008, p.133
167 Palmer, 1998, p.274
168 The reasons for and process by which Christian sacred writings were adopted has been explored by Wimbush (2000).
169 On the beginnings of the Black Church, see Mitchell (2004). Black Freemasonry, though not directly under the umbrella of the Black Church, kept alive Western esoteric traditions with Jewish and Christian strains. On its use of the Bible, see Page (2003).
170 The African American Heritage Hymnal has a section in its index calling attention to hymns on the topic of "God's Hand in Nature," but does not have responsive readings or litanies on this theme (see Carpenter and Williams 2001: prefatory matter and indices). The recently released, and now standard, textbook by Floyd-Thomas et al. does not list nature in its glossary of terms (2007: 247-258).
171 Pinn's survey of selected belief systems within the African American religious landscape, though released more than a decade ago, remains instructive in this regard (1998).

172 The story of Charles Gilbert's escape, as recounted by Still, is illustrative of this (2007: 119-125).

173 The shaman from whom Frederick Douglass received an implement of power and the conjurer – Gullah Jack Pritchard – who supported the South Carolina slave uprising led by Denmark Vesey are two well-known examples.

174 See http://www.sing365.com/music/lyric.nsf/why-i-sing-the-blues-lyrics-b-bking/

175 On the use of poetry, fiction, and memoir in academic writing, I have been inspired by the works of Staley (1995), Jones (1998), Denzin (1997), and others.

176 On the poetics of the "messy text," see Denzin (1997: 224-227).

177 For an assessment of its underlying patterns and more common paraphernalia, see Anderson (2008: 7-14).

178 Throughout this essay, I have been mindful that certain aspects of our understanding of nature are contested, of fairly recent origin, and culturally contingent.

179 Here, I adopt Theophus Smith's designation (1994: 8), in recognizing that such an enterprise is in keeping with his suggestion that "conjuring" be seen "as an indigenous spirituality with implications for revitalizing the next generation of black studies and black theology" (1994: 12).Consider one small example as an illustration. In *Ethics of the Fathers* (a compilation of *aggadic* material attributed to Rabbis of the Mishna), one finds this statement attributed to Rabbi Shimon:One, who while walking along the way, reviewing his studies, breaks off from his study and says, "How beautiful is that tree! How beautiful is that plowed field!" Scripture regards him as if he has forfeited his soul. (*Avot* 3:7)In the Zohar (the foundational text of Kabbalah, attributed to the 2nd century rabbi, Shimon bar Yochai although probably of later provenance), the same Rabbi Shimon is shown in a very different light: Rabbi Shimon, Rabbi Elazar, Rabbi Abba, and Rabbi Yossi were sitting under the trees in the valley of the Sea of Galilee. Rabbi Shimon said: "How beautiful is the shade with which these trees protect us; Let us crown them with words of Torah!" (Zohar, Part II, Parashat Teruma, 127a) In one Midrash, Shimon finds nature profane, in the other, profound. To complicate matters further, commentaries on each passage sometimes support, sometimes subvert the plain meanings of the texts. What is left is radical indeterminacy and a big collection of handy prooftexts for many different notions of nature, indeed for practically *any* view of nature. In this paper I will discuss only one of these many views.

181 See *She'elot U'Teshuvot Ha*-Maharshal, Lublin, 1599, Question 33, 24a-28b, and She'elot U'Teshuvot Ha-Rama, Warsaw, 1883, Question 6, 7a-8a.

182 Judah Loew b. Bezalel, Netivot 'Olam, Prague, 1596, Torah, Ch. 14, p. 69.

183 Quoted in M. Zuriel, "Ha-Hagbalot 'al ha-'Isuk ha-Sihli shel ha-Adam le-fi MaHaRaL," Ha-Ma'ayan, V. 27, N. 1, 1986, p. 55.

184 Judah Loew b. Bezalel, Tiferet Yisrael (Jerusalem, 1978) p. 35. (First edition: Venice, 1599).

185 Spinoza, Baruch and G. H. R. Parkinson, ed., (2000). Ethics. Oxford ; New York, Oxford University Press, Preface.

[186] Already in the late 19th century, Jews were found at the forefront of German science, which was itself among the most advanced in the world. In Germany, patterns emerged for the first time that would repeat themselves over the next half century wherever Jews lived in large numbers. Jews flocked to universities, a great many of them to seek training in science and medicine. By 1891, German Jews were more than eight times as likely as Christians to seek higher education, head for head. In 1891, one of every six medical students in Germany was a Jew. The percentages of Jews in mathematics, chemistry and physics were greater still, and this at a time when these faculties were themselves expanding rapidly. Jews were a rapidly growing portion of a rapidly growing population of science students, and soon enough they were a rapidly growing portion of a rapidly growing population of scientists, as well. Of the 100 Nobels awarded between 1901, when the prize was inaugurated, and 1932, when Hitler took power, 33 went to German scientists and one quarter of these went to Jews. And Germany was only a beginning, what in time would come to seem like a prototype. The rise of Jews in science and medicine in Russia and eventually in the Soviet Union, was swifter and greater still. By the last decades of the nineteenth century, the number of Jews in universities, and in the professions for which universities prepared them, had grown so rapidly that Russia's bureaucrats, pundits and hoi polloi concluded that enough was enough, introducing rules limiting Jews in medical schools, engineering schools and technical schools. After the revolution, such quotas ended abruptly. By 1939, Jews were ten times as likely to complete university studies as the general population, and one of every three college age Soviet Jews was studying at a university. These Jews mostly went into scientific and technical fields. In the early 1950s eleven percent of all Soviet scientists were Jews, though Jews comprised only one and a half percent of the population. (Harap, 1957, 55) In 1959, Jews were more than thirteen times as likely to be "scientific workers" than other Soviets were. The top five Jewish occupations were engineer, physician, "scientific personnel," teacher (most often in math and science) and chief production and technical managers (and the first three occupations accounted for 28% of all Jewish employment). Four of the seven Soviet Nobelists in physics were Jews (as were over thirty percent of all Soviet prizewinners up to 1975). (Patai 1977, 341) What Jews had achieved in Soviet science was exceeded only by what American Jews had achieved an ocean away.And that achievement was astonishing by any measure. Thirty eight percent of American laureates in physics are from Jewish backgrounds.

[186] Forty-two percent of American Nobelists in physiology and medicine are. Twenty-eight percent of U.S. prize winners in chemistry. And 37% of National Medals of Science recipients. When last studied (three decades ago), twenty-six percent of physicists in America's best universities were Jews. One in five mathematicians, bacteriologists, biochemists, and physiologists were. Greater numbers of physicians are Jewish, and greater percentages of the faculty at leading medical schools are Jewish.

[186] Who's Who in American Jewry of 1938 went to great pains to include everyone who had turned a good buck in business, as well as every famous Jewish

entertainer, athlete, newspaper reporter, novelist or rabbi. Still, about one in six of the 8,477 listings were men and women of science (over half of these physicians).

[187] Consider this crass marker as an illustration: among the Jewish Nobelists, whose numbers reach the hundreds, only one (Robert Aumann, who won the economics prize in 2005) has been religiously observant. The first century of the Nobel Prize produced no Jewishly observant laureates.

[188] Hyman Levy, (1933). The universe of science. New York, London,, The Century Co., p. 189:

[189] Gruenberg, B. C. (1935). Science and the Public Mind. New York and London,, McGraw-Hill book company, p. 36.

[190] Gruenberg, Science and the Public Mind, p. 39.

[191] Greunberg, Science and the Public Mind, p. 40. See as well the following: 40 "We know that scientists, along with the other classes, have manifested extreme provincialism and crude prejudice on occasion; but we are disposed to attribute such manifestations not to whatever science these men may have assimilated, but to human weakness not yet remedied by science" (p. 40) and "The Germans are today officially shutting their eyes to syphilis because the Wassermann test and the arsenical treatment were developed by Jews; and anti-Catholics would breed their plants and animals without making use of Mendel's aid. We cannot, of course, blame the scientists of Germany for the ridiculous and tragic outcomes of the political dictatorship in that country; but we may well consider that unless the scientists in this country cooperate wholeheartedly we cannot hope to cultivate among the general public a point of view which would make similar consequences unlikely in the future." (p. 168)

[192] Feynman, Richard (1985). "Surely you're joking, Mr. Feynman!" : adventures of a curious character. New York, W.W. Norton, p. 285.

[193] Stephen Jay Gould (1999), Rocks of Age, New York, Ballantine Books, p. 4.

[194] Though it is worth repeating that this attitude remains one among many. One finds today Jews whose attitude towards nature is one of apathy (a point of view prevalent among some traditions of what is sometimes called "ultra-orthodox" Judaism). One finds Jews who look to nature to verify Torah. One finds views that see in nature an invitation to a dangerous paganism. These attitudes too deserve elucidation, but this is beyond the scope of this presentation.